PENGUIN BOOKS

JOURNEY THROUGH GENIUS

William Dunham is a Phi Beta Kappa graduate of the University of Pittsburgh. After receiving his Ph.D. from the Ohio State University in 1974, he joined the mathematics faculty at Hanover College in Indiana. He has directed a summer seminar funded by the National Endowment for the Humanities on the topic of "The Great Theorems of Mathematics in Historical Context."

Journey Through Genius

THE GREAT THEOREMS OF MATHEMATICS

WILLIAM DUNHAM

PENGUIN BOOKS

PENGUIN BOOKS
Published by the Penguin Group
Penguin Group (USA) Inc., 375 Hudson Street, New York, New York 10014, U.S.A.
Penguin Group (Canada), 90 Eglinton Avenue East, Suite 700, Toronto,
Ontario, Canada M4P 2Y3 (a division of Pearson Penguin Canada Inc.)
Penguin Books Ltd, 80 Strand, London WC2R 0RL, England
Penguin Ireland, 25 St Stephen's Green, Dublin 2, Ireland (a division of Penguin Books Ltd)
Penguin Group (Australia), 250 Camberwell Road, Camberwell,
Victoria 3124, Australia (a division of Pearson Australia Group Pty Ltd)
Penguin Books India Pvt Ltd, 11 Community Centre, Panchsheel Park,
New Delhi – 110 017, India
Penguin Group (NZ), 67 Apollo Drive, Rosedale, North Shore 0632, New Zealand
(a division of Pearson New Zealand Ltd)
Penguin Books (South Africa) (Pty) Ltd, 24 Sturdee Avenue, Rosebank,
Johannesburg 2196, South Africa

Penguin Books Ltd, Registered Offices: 80 Strand, London WC2R 0RL, England

First published in the United States of America by
John Wiley & Sons, Inc., 1990
Published in Penguin Books 1991

40 39 38 37 36

THE LIBRARY OF CONGRESS HAS CATALOGUED THE HARDCOVER AS FOLLOWS:
Dunham, William, 1947–
Journey through genius: the great theorems of mathematics / William Dunham.
p. cm.
Includes bibliographical references (p. 287).
ISBN 0-471-50030-5(hc.)
ISBN 978-0-14-014739-1 (pbk.)
1. Mathematics—History. 2. Mathematicians—Biography.
I. Title.
QA21.D78 1990
510'.9—dc20 89–27366

Printed in the United States of America

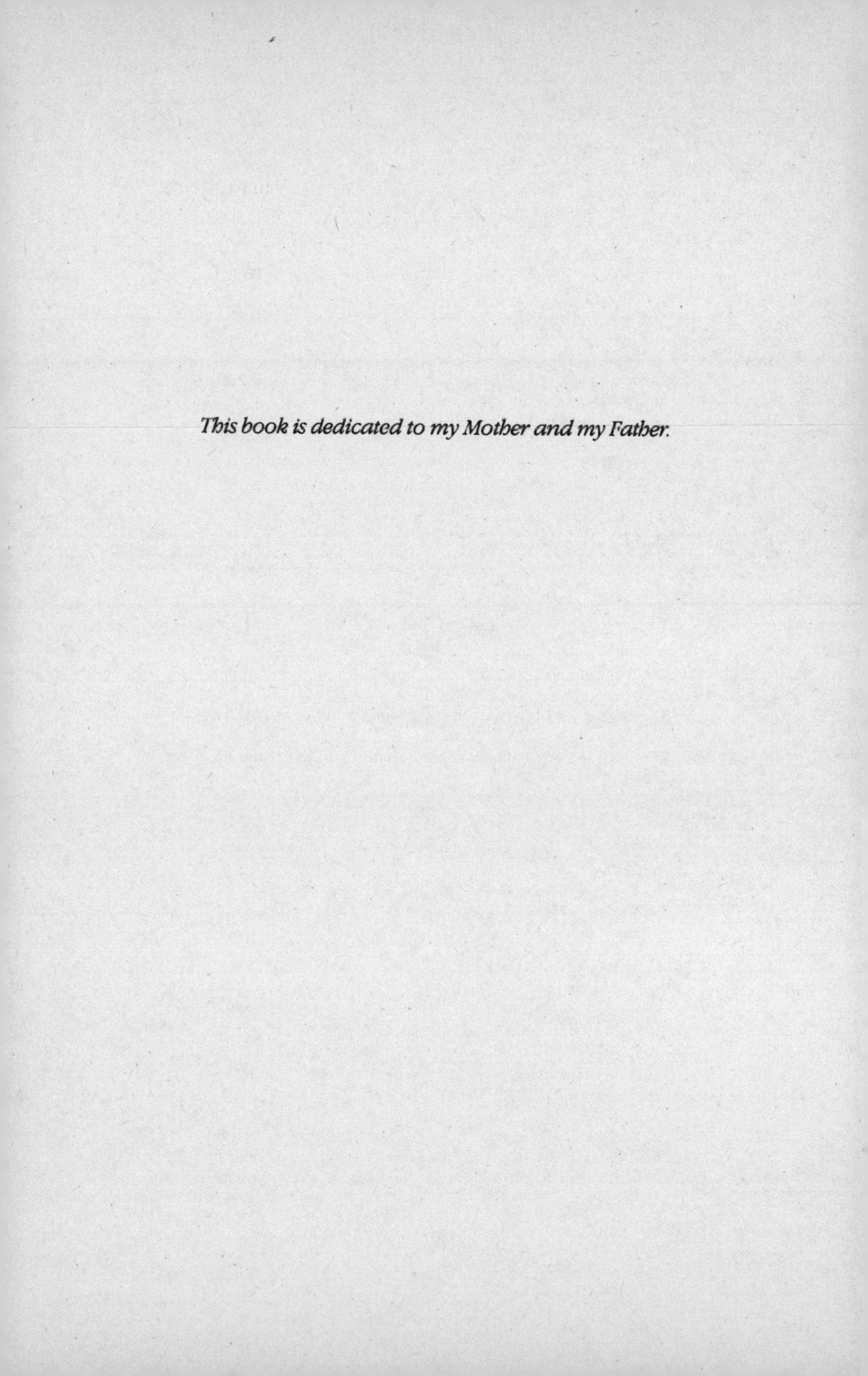

This book is dedicated to my Mother and my Father.

Preface

In his autobiography, Bertrand Russell recalled the crisis of his youth:

> There was a footpath leading across fields to New Southgate, and I used to go there alone to watch the sunset and contemplate suicide. I did not, however, commit suicide, because I wished to know more of mathematics.

Admittedly, few people find such absolute salvation in mathematics, but many appreciate its power and, more critically, its beauty. This book is designed for those who would like to probe a bit more deeply into the long and glorious history of mathematics.

For disciplines as diverse as literature, music, and art, there is a tradition of examining masterpieces—the "great novels," the "great symphonies," the "great paintings"—as the fittest and most illuminating objects of study. Books are written and courses are taught on precisely these topics in order to acquaint us with some of the creative milestones of the discipline and with the men and women who produced them.

The present book offers an analogous approach to mathematics, where the creative unit is not the novel or symphony, but the theorem. Consequently, this is not a typical math book in that it does not provide a step-by-step development of some branch of the subject. Nor does it stress the applicability of mathematics in determining planetary orbits, in understanding the world of computers, or, for that matter, in balancing your checkbook. Mathematics, of course, has been spectacularly successful in such applied undertakings. But it was not its worldly utility that led Euclid or Archimedes or Georg Cantor to devote so much of their energy and genius to mathematics. These individuals did not feel compelled to justify their work with utilitarian applications any more than Shakespeare had to apologize for writing love sonnets instead of cookbooks or Van Gogh had to apologize for painting canvases instead of billboards.

In this book I shall explore a handful of the most important proofs—and the most ingenious logical arguments—from the history of mathematics, with emphasis on why the theorems were significant and how the mathematician resolved, once and for all, the pressing logical issue. Each chapter of *Journey Through Genius* has three primary components:

The first is its *historical* emphasis. The "great theorems" on the pages ahead span more than 2300 years of human history. Before discussing a particular result, I shall set the scene by describing the state of mathematics, and perhaps the state of the world generally, prior to the theorem. Like everything else, mathematics is created within the context of history, and it is of interest to place Cardano's solution of the cubic two years after the publication of Copernicus's heliocentric theory and two years before the death of England's Henry VIII, or to emphasize the impact of the Restoration upon Cambridge University when a young scholar named Isaac Newton entered it in 1661.

The second component is the *biographical*. Mathematics is the product of real, flesh-and-blood human beings whose lives may reflect the inspirational, the tragic, or the bizarre. The theorems contained here represent the work of a number of individuals, ranging from the gregarious Leonhard Euler to the pugnacious Johann Bernoulli to that most worldly of Renaissance characters, Gerolamo Cardano. Understanding something of the lives of these diverse individuals can only enhance an appreciation of their work.

The final component, and the primary focus of the book, is the creativity evident in these "mathematical masterpieces." Just as one could not hope to understand a great novel without reading it, or to appreciate a great painting without seeing it, so one cannot really come to grips with a great mathematical theorem without a careful, step-by-step look at the proof. To acquire such an understanding requires a good bit of concentration and effort, and the chapters to follow are meant to serve as a guide in that undertaking.

There is a remarkable permanence about these mathematical landmarks. In other disciplines, the fads of today become the forgotten discards of tomorrow. A little over a century ago, Sir Walter Scott was among the most esteemed writers in English literature; today, he is regarded considerably less enthusiastically. In the twentieth century, superstars come and go with breathtaking speed, and ideas that seem destined to change the world often end up on the intellectual scrap heap.

Mathematics, to be sure, is also subject to changes of taste. But a theorem, correctly proved within the severe constraints of logic, is a theorem forever. Euclid's proof of the Pythagorean theorem from 300 B.C. has lost none of its beauty or validity with the passage of time. By contrast, the astronomical theories or medical practices of Alexandrian Greece have long since become archaic, slightly amusing examples of primitive science. The nineteenth-century mathematician Hermann Hankel said it best:

> In most sciences one generation tears down what another has built, and what one has established another undoes. In mathematics alone each generation adds a new story to the old structure.

In this sense, as we examine the timeless mathematics of great mathematicians, we come to understand Oliver Heaviside's wonderfully apt observation: "Logic can be patient, for it is eternal."

A number of factors have gone into the selection of these few theorems to represent the best of mathematics. As noted, my chief consideration was to find arguments that were particularly insightful or ingenious. This, of course, introduces an element of personal taste, and I recognize that a different author would certainly generate a different list of great theorems. That aside, it is an extraordinary experience to behold, first-hand, the mathematician gliding through clever deductions and making the seemingly incomprehensible become clear. It has been said that *talent* is doing easily what others find difficult, but that *genius* is doing easily what others find impossible. As will be evident, there is much genius displayed on the pages ahead. Here are genuine classics—the *Mona Lisas* or *Hamlets* of mathematics.

But other considerations influenced the choice of theorems. For one, I wanted to include samples from history's leading mathematicians. It was a must, for instance, to have selections from Euclid, Archimedes, Newton, and Euler. To overlook such figures would be like studying art history without mentioning the work of Rembrandt or Cézanne.

Further, for the sake of variety, I have sampled different branches of mathematics. The propositions in the book come from the realms of plane geometry, algebra, number theory, analysis, and the theory of sets. The variety of these topics, and the occasional links and interplays among them, may add a note of freshness to this work.

I also wanted to present *important* mathematical theorems, rather than merely clever little tricks or puzzles. Indeed, most of the results in the book either resolved long-standing problems in mathematics, or generated even more profound questions for the future, or both. At the end of each chapter is an Epilogue, usually addressing an issue raised by the great theorem and following it as it echoes down through the history of mathematics.

Then there is the question of level of difficulty. Obviously, mathematics has many great landmarks whose depth and complexity render them incomprehensible to all but experts. It would be foolish to include such results in a book aimed at the general, scientifically literate reader. The theorems that follow require only the tools of algebra and geometry, of the sort one acquires in a few high school courses. The two exceptions are a brief use of the sine curve from trigonometry in discussing the work of Euler in Chapter 9 and an application of elementary integral calculus in the work of Newton in Chapter 7; many readers may already be acquainted with these topics, and for those who are not, there is a bit of explanation to smooth over the difficulty.

I should stress that this is not a scholarly tome. There are certainly questions of great mathematical or historical subtlety that cannot be addressed in a work of this kind. While I have tried to avoid including false or historically inaccurate material, this was simply not the time nor place to investigate all facets of all issues. This book, after all, is meant for the popular, not the scientific, press.

Along these lines, I must add a word about the authenticity of the proofs. In preparing the book, I have found it impossible to avoid the need for some compromise between the authors' original notation, terminology, and logical strategy and the requirement that the mathematical material be understandable to the modern reader. A complete adherence to the originals would make some of these results very difficult to comprehend; yet a significant deviation from the originals would conflict with my historical objectives. In general, I have tried to retain virtually all of the spirit, and a good bit of the detail, of the original theorems. The modifications I have introduced seem to me to be no more serious, than, say, performing Mozart on modern instruments.

And so, we are about to begin our journey through two millennia of mathematical landmarks. These results, old as they are, retain a freshness and display a sparkling virtuosity even after so many centuries. I hope that the reader will be able to understand these proofs and to recognize what made them great. For those who succeed in this venture, I expect there will be not only a sense of awe that comes from appreciating the greatness of others, but also a sense of personal satisfaction that one can, indeed, comprehend the works of a master.

W. Dunham
Columbus, Ohio

Acknowledgments

I am indebted to a number of agencies and individuals for their efforts on my behalf in the preparation of this book. First, I must acknowledge grants from both the private and the public sectors that were invaluable: a 1983 Summer Stipend from the Lilly Endowment, Inc., and the funding of a 1988 Summer Seminar titled "The Great Theorems of Mathematics in Historical Context" by the National Endowment for the Humanities. The support of Lilly Endowment and NEH allowed my previously unfocused interest in the history of mathematics to take shape in the form of courses at Hanover College and Ohio State.

To The Ohio State University, and particularly to its Department of Mathematics, go my sincere thanks for their warm hospitality while, as a visiting faculty member in their midst, I was writing this book. I shall always appreciate the kindness of Department Chair Joseph Ferrar and of the Leitzels—Joan and Jim—who were unfailingly helpful and supportive during my two-year visit.

Many individuals also contributed to this work. Thanks go to Ruth Evans, my favorite librarian, who introduced me to the collections of pre-1900 mathematical documents during my 1980 sabbatical; to Steven

Tigner and Michael Hall of NEH for their good advice about the summer seminar that preceded this book; to Carol Dunham for her enthusiasm and encouragement; to Amy Edwards and Jill Baumer-Piña of Ohio State for introducing me to the finer points of Macintosh word processing; to my Wiley editors Katherine Schowalter, Laura Lewin and Steve Ross for their tolerance of a first-time author; to V. Frederick Rickey of Bowling Green State University, one of the nation's most effective spokespersons for the idea that mathematics—like other disciplines—has a history that must not be ignored; to Barry A. Cipra and to Russell Howell of Westmont College for their thorough and very helpful review of the manuscript; and to Jonathan Smith of Hanover College for his editorial comments in the final stages before publication.

Most of all, my gratitude goes to Penny Dunham, who prepared the illustrations for the book and made many valuable suggestions about its contents. Penny is an extraordinary teacher of mathematics, an irreplaceable colleague in our joint NEH seminar, a supporter, advisor, spouse, and the best friend imaginable.

Finally, an extra special thank you to Brendan and Shannon, who are simply the greatest.

Contents

Chapter 1

Hippocrates' Quadrature of the Lune

(ca. 440 B.C.)

The Appearance of Demonstrative Mathematics

Our knowledge of the very early development of mathematics is largely speculative, pieced together from archaeological fragments, architectural remains, and educated guesses. Clearly, with the invention of agriculture in the years 15,000–10,000 B.C., humans had to address, in at least a rudimentary fashion, the two most fundamental concepts of mathematics: multiplicity and space. The notion of multiplicity, or "number," would arise when counting sheep or distributing crops; over the centuries, refined and extended by generations of scholars, these ideas evolved into arithmetic and later into algebra. The first farmers likewise would have needed insight into spatial relationships, primarily in regard to the areas of fields and pastures; such insights, carried down through history, became geometry. From the beginnings of civilization, these two great branches of mathematics—arithmetic and geometry—would have coexisted in primitive form.

This coexistence has not always been a harmonious one. A continuing feature of the history of mathematics has been the prevailing tension between the arithmetic and the geometric. There have been times when one branch has overshadowed the other and when one has been regarded as logically superior to its more suspect counterpart. Then a new discovery, a new point of view, would turn the tables. It may come as a surprise that mathematics, like art or music or literature, has been subject to such trends in the course of its long and illustrious history.

We find clear signs of mathematical development in the civilization of ancient Egypt. For the Egyptians, the emphasis was on the practical side of mathematics as a facilitator of trade, agriculture, and the other increasingly complex aspects of everyday life. Archaeological records indicate that by 2000 B.C. the Egyptians had a primitive numeral system as well as some geometric ideas about triangles, pyramids, and the like. There is a tradition, for instance, that Egyptian architects used a clever device for making right angles. They would tie 12 equally long segments of rope into a loop, as shown in Figure 1.1. Stretching five consecutive segments in a straight line from *B* to *C* and then pulling the rope taut at *A*, they thus formed a rigid triangle with a right angle *BAC*. This configuration, laid upon the ground, allowed the workers to construct a perfect right angle at the corner of a pyramid, temple, or other building.

Implicit in this construction is an understanding of the Pythagorean relationship of right triangles. That is, the Egyptians seemed to know that a triangle with sides of length 3, 4, and 5 must contain a right angle. Of course, $3^2 + 4^2 = 9 + 16 = 25 = 5^2$, and so we catch an early glimpse of one of the most important relationships in all of mathematics (see Figure 1.2).

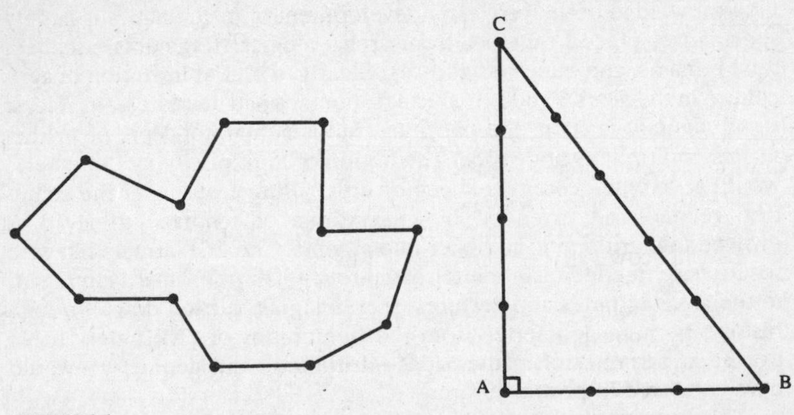

FIGURE 1.1

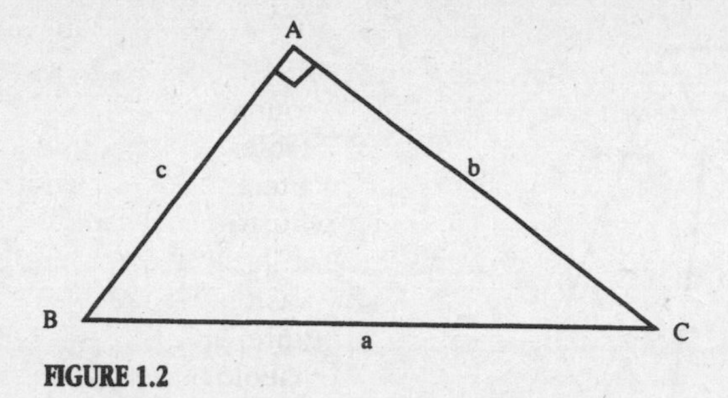

FIGURE 1.2

Technically, this Egyptian insight was not a case of the Pythagorean theorem itself, which states, "If $\triangle BAC$ is a right triangle, then $a^2 = b^2 + c^2$." Rather, it was an example of the *converse* of the Pythagorean theorem: "If $a^2 = b^2 + c^2$, then $\triangle BAC$ is a right triangle." That is, for a proposition of the form "If *P*, then *Q*," the related statement "If *Q*, then *P*" is called the proposition's "converse." As we shall see, a perfectly true statement may have a false converse, but in the case of the famous Pythagorean theorem, both the proposition and its converse are valid. In fact, these will be the "great theorems" in the next chapter.

Although the Egyptians seemed to have some insight into the geometry of 3-4-5 right triangles, it is doubtful they possessed the broader understanding that, for instance, a 5-12-13 triangle or a 65-72-97 triangle likewise contains a right angle (since in each case $a^2 = b^2 + c^2$). More critically, the Egyptians gave no indication of how they might *prove* this relationship. Perhaps they had some logical argument to support their observation about 3-4-5 triangles; perhaps they hit upon it purely by trial and error. In any case, the notion of proving a general mathematical result by a carefully crafted logical argument is nowhere to be found in Egyptian writings.

The following example of Egyptian mathematics may be illuminating: it is their approach to finding the volume of a truncated square pyramid—that is, a square pyramid with its top lopped off by a plane parallel to the base (see Figure 1.3). Such a solid is today called the frustum of a pyramid. The technique for finding its volume appears in the so-called "Moscow Papyrus" from 1850 B.C.:

> If you are told: A truncated pyramid of 6 for the vertical height by 4 on the base by 2 on the top. You are to square this 4, result 16. You are to double 4, result 8. You are to square 2, result 4. You are to add the 16, the 8, and the 4, result 28. You are to take a third of 6, result 2. You are to take 28 twice, result 56. See, it is 56. You will find it right.

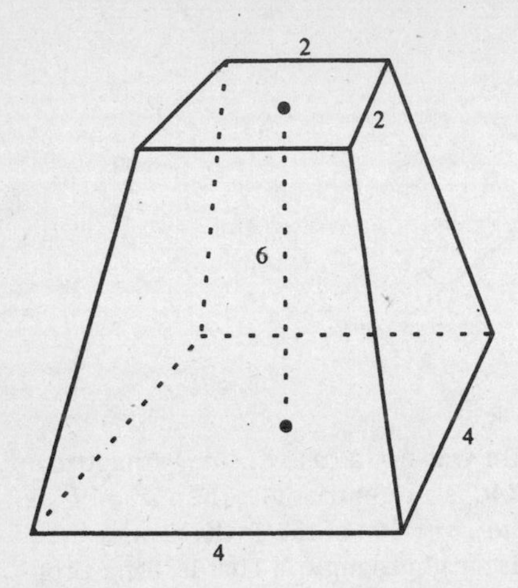

FIGURE 1.3

This is a most remarkable prescription, which indeed yields the correct answer for the frustum's volume. Notice, however, what it does not do. It does not give a general formula to cover frusta of other dimensions. Egyptians would have to generalize from this particular case in order to determine the volume of a different-sized frustum, a process that could be a bit confusing. Far simpler and more concise is our modern formula

$$V = \frac{1}{3}h(a^2 + ab + b^2)$$

where *a* is the side of the square on the bottom, *b* is the side of the square on the top, and *h* is the frustum's height. Worse, there was no indication of *why* this Egyptian recipe provided the correct answer. Instead, a simple "You will find it right" sufficed.

It is probably dangerous to draw sweeping conclusions from a particular example, yet historians have noted that a dogmatic approach to mathematics was certainly in keeping with the authoritarian society that was pharaonic Egypt. Inhabitants of that ancient land were conditioned to give unquestioned obedience to their rulers. By analogy, when presented with an authoritative mathematical technique that concluded "You will find it right," Egyptian subjects were hardly likely to demand a more thorough explanation of why it worked. In the land of the Pharaoh, you did what you were told, whether in erecting a colossal temple or in solving a math problem. Those adamantly questioning the system would end up as mummies before their time.

Another great ancient civilization—or, more precisely, civilizations—flourished in Mesopotamia and produced mathematics significantly more advanced than that of Egypt. The Babylonians, for instance, solved fairly sophisticated problems with a definite algebraic character, and the existence of a clay tablet called Plimpton 322, dated roughly between 1900 and 1600 B.C., shows that they definitely understood the Pythagorean theorem in far more depth than their Egyptian counterparts; that is, the Babylonians recognized that a 5-12-13 triangle or a 65-72-97 triangle (and many more) was right. In addition, they developed a sophisticated place system for their numerals. We, of course, are accustomed to a base-10 numeral system, obviously derived from the 10 fingers of the human hand, so it may seem a bit odd that the Babylonians chose a base-60 system. While no one speculates that these ancient people had 60 fingers, their choice of base can still be seen in our measurement of time (60 seconds per minute) and angles ($6 \times 60° = 360°$ in a circle).

But for all of their achievements, the Mesopotamians likewise addressed only the question of "how" while avoiding the much more significant issue of "why." Those seeking the appearance of a demonstrative mathematics—a theoretical, deductive system in which emphasis was placed upon *proving* critical relationships—would have to look to a later time and a different place.

The time was the first millennium B.C., and the place was the Aegean coasts of Asia Minor and Greece. Here there arose one of the most significant civilizations of history, whose extraordinary achievements would forever influence the course of western culture. Engaged in a thriving commerce, both within their own lands and across the Mediterranean, the Greeks developed into a mobile, adventuresome people, relatively prosperous and sophisticated, and considerably more independent in thought and action than the western world had seen before. These curious, free-thinking merchants were much less likely to submit meekly to authority. Indeed, with the development of Greek democracy, the citizens *became* the authority (although it must be stressed that citizenship in the classical world was very narrowly defined). To such individuals, everything was open to debate and analysis, and ideas were not about to be accepted with a passive, unquestioning obedience.

By 400 B.C., this remarkable civilization could already boast a rich, some would say unsurpassed, intellectual heritage. The epic poet Homer, the historians Herodotus and Thucydides, the dramatists Aeschylus, Sophocles, and Euripides, the politician Pericles, and the philosopher Socrates—these individuals had all left their marks as the fourth century B.C. began. Inhabitants of the modern world, where fame can fade so quickly, may find it astonishing that these names have endured gloriously for over 2000 years. To this day, we admire their boldness in subjecting Nature and the human condition to the penetrating light of reason. Granted, it was reason still contaminated by large doses of superstition and ignorance, but the Greek thinkers were profoundly successful. If their conclusions were not always correct, the Greeks nonetheless sensed that theirs was the path that would lead from a barbarous past to an undreamed-of future. The term "awakening" is often used in describing this special moment in history, and it is apt. Humankind was indeed arising from the slumber of thousands of centuries to confront this strange, mysterious world with Nature's most potent weapon—the human mind.

Such was certainly the case with mathematics. Around 600 B.C. in the town of Miletus on the western coast of Asia Minor, there lived the great Thales (ca. 640–ca. 546 B.C.), one of the so-called "Seven Wise Men" of antiquity. Thales of Miletus is generally credited with being the father of demonstrative mathematics, the first scholar who supplied the "why" along with the "how." As such, he is the earliest known mathematician.

We have very little hard evidence about his life. Indeed, he emerges from the mists of the past as a pseudo-mythical figure, and it is anybody's guess as to the truth of the exploits and discoveries attributed to him. Looking back seven centuries, the biographer Plutarch (A.D. 46–120) wrote that ". . . at that time Thales alone had raised philosophy above mere practice into speculation." A noted mathematician and astronomer who somehow predicted the solar eclipse in 585 B.C., Thales, like the stereotypical scientist, was chronically absent-minded and incessantly preoccupied—according to legend, he once was strolling along, gazing upward at his beloved stars, when he tumbled into an open well.

His "fatherhood" of demonstrative mathematics notwithstanding, Thales never married. When Solon, a contemporary, asked why, Thales arranged a cruel ruse whereby a messenger brought Solon news of his son's death. According to Plutarch, Solon then

> . . . began to beat his head and to do and say all that is usual with men in transports of grief. But Thales took his hand, and, with a smile, said, "These things, Solon, keep me from marriage and rearing children, which are too great for even your constancy to support; however, be not concerned at the report, for it is a fiction."

Clearly, Thales was not the kindest of people. A similar impression emerges from the story of a farmer who routinely tied heavy bags of salt on the back of his donkey when driving the beast to market. The clever animal quickly learned to roll over while fording a particular stream, thereby dissolving much of the salt and making his burden far lighter. Exasperated, the farmer went to Thales for advice, and Thales recommended that on the next trip to market the farmer load the donkey with sponges.

It was certainly not kindness to man or beast that earned Thales his high reputation in mathematics. Rather, it was his insistence that geometric statements not be accepted simply because of their intuitive plausibility; instead they had to be subjected to rigorous, logical proof. This is no small legacy to leave the discipline of mathematics.

What, precisely, are some of his theorems? Tradition holds that it was Thales who first *proved* the following geometric results:

- Vertical angles are equal.
- The angle sum of a triangle equals two right angles.
- The base angles of an isosceles triangle are equal.
- An angle inscribed in a semicircle is a right angle.

In none of these cases do we have any record of his proofs, but we can speculate on their nature. For instance, consider the last proposition above. The proof given below is taken from Euclid's *Elements*, Book III, Proposition 31, but it is simple and direct enough to be a prime candidate for Thales' own.

THEOREM An angle inscribed in a semicircle is a right angle.

PROOF Let a semicircle be drawn with center O and diameter BC, and choose any point A on the semicircle (Figure 1.4). We must prove that $\angle BAC$ is right. Draw line OA and consider $\triangle AOB$. Since OB and OA are radii of the semicircle, they have the same length, and so $\triangle AOB$ is isosceles. Hence, as Thales had previously proved, $\angle ABO$ and $\angle BAO$ are equal (or, in modern terminology, congruent); call them both α. Like-

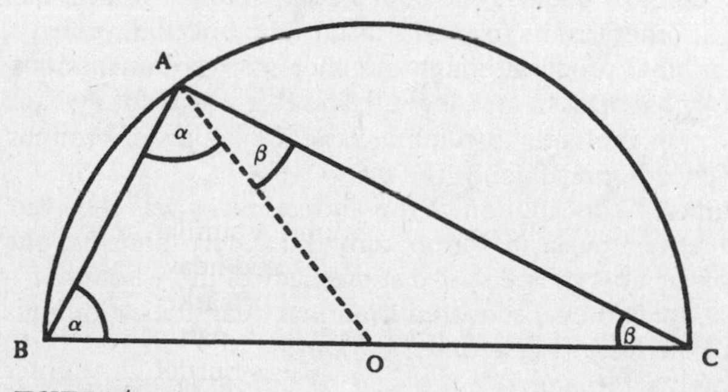

FIGURE 1.4

wise, in $\triangle AOC$, OA and OC have the same length, and so $\angle OAC = \angle OCA$; call them both β. But, from the large triangle BAC, we see that

$$\begin{aligned} 2 \text{ right angles} &= \angle ABC + \angle ACB + \angle BAC \\ &= \alpha + \beta + (\alpha + \beta) \\ &= 2\alpha + 2\beta = 2(\alpha + \beta) \end{aligned}$$

Hence, one right angle $= \frac{1}{2}[2 \text{ right angles}] = \frac{1}{2}[2(\alpha + \beta)] = \alpha + \beta = \angle BAC$. This is exactly what we were to prove.

Q.E.D.

(*Note:* It has become customary, upon the completion of a proof, to insert the letters "Q.E.D.," which abbreviate the Latin *Quod erat demonstrandum* [Which was to be proved]. This alerts the reader to the fact that the argument is over and we are about to set off in new directions.).

After Thales, the next major figure in Greek mathematics was Pythagoras. Born in Samos around 572 B.C., Pythagoras lived and worked in the eastern Aegean, even, according to some legends, studying with the great Thales himself. But when the tyrant Polycrates assumed power in this region, Pythagoras fled to the Greek town of Crotona in southern Italy, where he founded a scholarly society now known as the Pythagorean brotherhood. In their contemplation of the world about them, the Pythagoreans recognized the special role of "whole number" as the critical foundation of all natural phenomena. Whether in music, or astronomy, or philosophy, the central position of "number" was everywhere evident. The modern notion that the physical world can be understood by "mathematization" owes more than a little to this Pythagorean viewpoint.

In the world of mathematics proper, the Pythagoreans gave us two great discoveries. One, of course, was the incomparable Pythagorean theorem. As with all other results from this distant time period, we have no record of the original proof, although the ancients were unanimous in attributing it to Pythagoras. In fact, legend says that a grateful Pythagoras sacrificed an ox to the gods to celebrate the joy his proof brought to all concerned (except, presumably, the ox).

The other significant contribution of the Pythagoreans was received with considerably less enthusiasm, for not only did it defy intuition, but it also struck a blow against the pervasive supremacy of the whole number. In modern parlance, they discovered irrational quantities, although their approach had the following geometric flavor:

Two line segments, AB and CD, are said to be *commensurable* if there exists a smaller segment EF that goes evenly into both AB and CD.

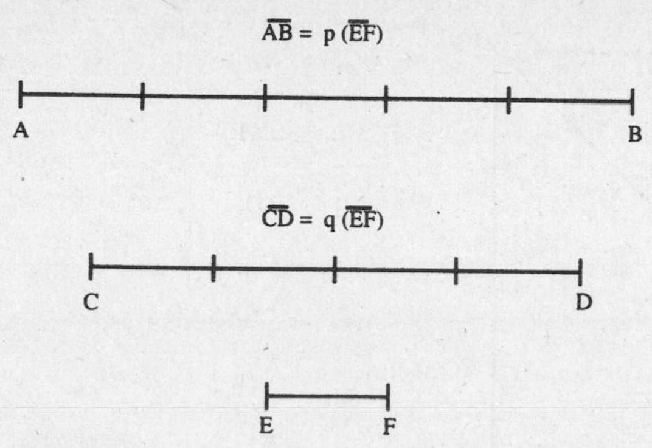

FIGURE 1.5

That is, for some whole numbers p and q, AB is composed of p segments congruent to EF while CD is composed of q such segments (Figure 1.5). Consequently, $\overline{AB}/\overline{CD} = p(\overline{EF})/q(\overline{EF}) = p/q$. (Here we are using the notation $\overline{AB}$ to stand for the length of segment AB). Since p/q is the *ratio* of two positive integers, we say that the ratio of the lengths of commensurable segments is a "rational" number.

Intuitively, the Pythagoreans felt that *any* two magnitudes are commensurable. Given two line segments, it seemed preposterous to doubt the existence of another segment EF dividing evenly into both, even if it took an extremely tiny EF to do the job. The presumed commensurability of segments was critical to the Pythagoreans, not only because they used this idea in their proofs about similar triangles but also because it seemed to support their philosophical stance on the central role of whole numbers.

However, tradition credits the Pythagorean Hippasus with discovering that the side of a square and its diagonal (GH and GI in Figure 1.6) are not commensurable. That is, no matter how small one goes, there is no magnitude EF dividing *evenly* into both the square's side and its diagonal.

This discovery had a number of profound consequences. Obviously, it shattered those Pythagorean proofs that rested upon the supposed commensurability of all segments. It would be almost two centuries before the mathematician Eudoxus found a way to patch up the theory of similar triangles by devising alternative proofs that did not rely upon the concept of commensurability. Secondly, it had an unsettling impact upon the supremacy of whole numbers, for if not all quantities were commensurable, then whole numbers were somehow inadequate to represent the ratios of all geometric lengths. Consequently, the discovery

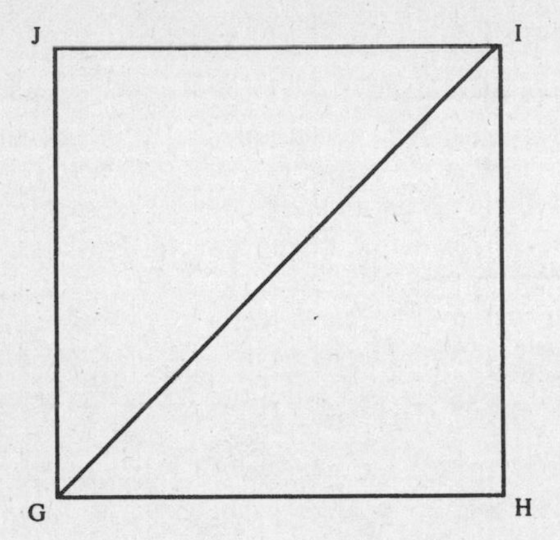

FIGURE 1.6

firmly established the superiority of geometry over arithmetic in all subsequent Greek mathematics. In the Figure 1.6, for instance, the side and diagonal of the square are beyond suspicion as *geometric* objects. But, as *numbers,* they presented a major problem. For, if we imagine that the side of the square above has length 1, then the Pythagorean theorem tells us that the length of the diagonal is $\sqrt{2}$; and, since side and diagonal are not commensurable, we see that $\sqrt{2}$ cannot be written as a rational number of the form p/q. Numerically, then, $\sqrt{2}$ is an "irrational," whose arithmetic character is quite mysterious. Far better, thought the Greeks, to avoid the numerical approach altogether and concentrate on magnitudes simply as geometric entities. This preference for geometry over arithmetic would dominate a thousand years of Greek mathematics.

A final result of the discovery of irrationals was that the Pythagoreans, incensed at all the trouble Hippasus had caused, supposedly took him far out upon the Mediterranean and tossed him overboard to his death. If true, the story indicates the dangers inherent in free thinking, even in the relatively austere discipline of mathematics.

Thales and Pythagoras, while prominent in legend and tradition, are obscure, shadowy figures from the distant past. Our next individual, Hippocrates of Chios (ca. 440 B.C.) is a little more solid. In fact, it is to him that we attribute the earliest mathematical proof that has survived in reasonably authentic form. This will be the subject of our first great theorem.

Hippocrates was born on the island of Chios sometime in the fifth century B.C. This was, of course, the same region that produced his illustrious predecessors mentioned earlier. (Note in passing that Chios is not far from the island of Cos, where another "Hippocrates" was born about this time; it was Hippocrates of Cos—not our Hippocrates—who became the father of Greek medicine and originator of the physicians' Hippocratic oath.)

Of the mathematical Hippocrates, we have scant biographical information. Aristotle wrote that, while a talented geometer, he "... seems in other respects to have been stupid and lacking in sense." This is an early example of the stereotype of the mathematician as being somewhat overwhelmed by the demands of everyday life. Legend has it that Hippocrates earned this reputation after being defrauded of his fortune by pirates, who apparently took him for an easy mark. Needing to make a financial recovery, he traveled to Athens and began teaching, thus becoming him one of the few individuals ever to enter the teaching profession for its financial rewards.

In any case, Hippocrates is remembered for two signal contributions to geometry. One was his composition of the first *Elements,* that is, the first exposition developing the theorems of geometry precisely and logically from a few given axioms or postulates. At least, he is credited with such a work, for nothing remains of it today. Whatever merits his book had were to be eclipsed, over a century later, by the brilliant *Elements* of Euclid, which essentially rendered Hippocrates' writings obsolete. Still, there is reason to believe that Euclid borrowed from his predecessor, and thus we owe much to Hippocrates for his great, if lost, treatise.

The other significant Hippocratean contribution—his quadrature of the lune—fortunately has survived, although admittedly its survival is tenuous and indirect. We do not have Hippocrates' own work, but Eudemus' account of it from around 335 B.C., and even here the situation is murky, because we do not really have Eudemus' account either. Rather, we have a summary by Simplicius from A.D. 530 that discussed the writings of Eudemus, who, in turn, had summarized the work of Hippocrates. The fact that the span between Simplicius and Hippocrates is almost a thousand years—roughly the time between us and Leif Erikson—indicates the immense difficulty historians face when considering the mathematics of the ancients. Nonetheless, there is no reason to doubt the general authenticity of the work in question.

Some Remarks on Quadrature

Before examining Hippocrates' lunes, we need to address the notion of "quadrature." It is obvious that the ancient Greeks were enthralled by the symmetries, the visual beauty, and the subtle logical structure of geometry. Particularly intriguing was the manner in which the simple and elementary could serve as foundation for the complex and intricate. This will become quite apparent in the next chapter as we follow Euclid through the development of some very sophisticated geometric propositions beginning with just a few basic axioms and postulates.

This enchantment with building the complex from the simple was also evident in the Greeks' geometric constructions. For them, the rules of the game required that all constructions be done only with compass and (unmarked) straightedge. These two fairly unsophisticated tools—allowing the geometer to produce the most perfect, uniform one-dimensional figure (the straight line) and the most perfect, uniform two-dimensional figure (the circle)—must have appealed to the Greek sensibilities for order, simplicity, and beauty. Moreover, these constructions were within reach of the technology of the day in a way that, for instance, constructing a parabola was not. Perhaps it is accurate to suggest that the aesthetic appeal of the straight line and circle reinforced the central position of straightedge and compass as geometric tools while, conversely and simultaneously, the physical availability of these tools enhanced the role to be played by straight lines and circles in the geometry of the Greeks.

The ancient mathematicians were consequently committed to, and limited by, the output of these tools. As we shall see, even the seemingly unsophisticated compass and straightedge can produce, in the hands of ingenious geometers, a rich and varied set of constructions, from the bisection of lines and angles, to the drawing of parallels and perpendiculars, to the creation of regular polygons of great beauty. But a considerably more challenging problem in the fifth century B.C. was that of the quadrature or squaring of a plane figure. To be precise:

□ The *quadrature* (or squaring) of a plane figure is the construction—using only compass and straightedge—of a square having area equal to that of the original plane figure. If the quadrature of a plane figure can be accomplished, we say that the figure is *quadrable* (or squarable).

That the quadrature problem appealed to the Greeks should come as no surprise. From a purely practical viewpoint, the determination of the area of an irregularly shaped figure is, of course, no easy matter. If such a figure could be replaced by an equivalent square, then determining the original area would have been reduced to the trivial matter of finding the area of that square.

Undoubtedly the Greeks' fascination with quadrature went far beyond the practical. For, if successfully accomplished, quadrature would impose the symmetric regularity of the square onto the asymmetric irregularity of an arbitrary plane figure. To those who sought a natural world governed by reason and order, there was much appeal in the process of replacing the asymmetric by the symmetric, the imperfect by the perfect, the irrational by the rational. In this sense, quadrature represented not only the triumph of human reason, but also the inherent simplicity and beauty of the universe itself.

Devising quadratures was thus a particularly fascinating problem for Greek mathematicians, and they produced clever geometric constructions to that end. As is often the case in mathematics, solutions can be approached in stages, by first squaring a reasonably "tame" figure and moving from there to the quadrature of more irregular, bizarre ones. The key initial step in this process is the quadrature of the rectangle, the procedure for which appears as Proposition 14 of Book II of Euclid's *Elements,* although it was surely known well before Euclid. We begin with this.

STEP 1 Quadrature of the rectangle (Figure 1.7)

Let *BCDE* be an arbitrary rectangle. We must construct, with compass and straightedge only, a square having area equal to that of *BCDE.* With the straightedge, extend line *BE* to the right, and use the compass to mark off segment *EF* with length equal to that of *ED*—that is, $\overline{EF} = \overline{ED}$. Next, bisect *BF* at *G* (an easy compass and straightedge construction), and with center *G* and radius $\overline{BG} = \overline{FG}$, describe a semicircle as shown. Finally, at *E*, construct line *EH* perpendicular to *BF*, where *H* is the point of intersection of the perpendicular and the semicircle, and from there construct square *EKLH.*

We now claim that the shaded square having side of length $\overline{EH}$—a figure we have just *constructed*—has area equal to that of the original rectangle *BCDE.*

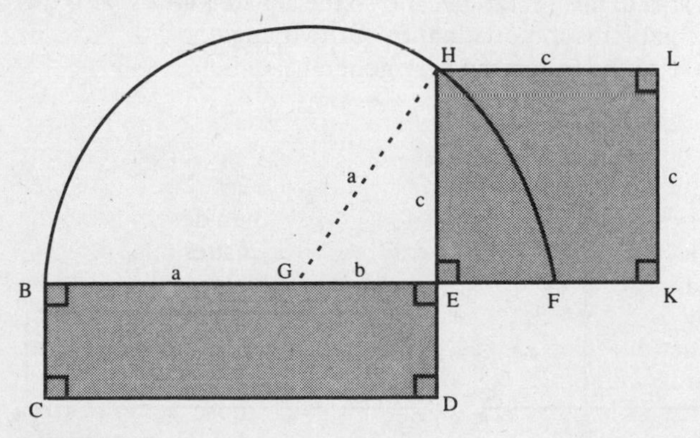

FIGURE 1.7

To verify this claim requires a bit of effort. For notational convenience, let a, b, and c be the lengths of segments HG, EG, and EH, respectively. Since $\triangle GEH$ is a right triangle by construction, the Pythagorean theorem gives us $a^2 = b^2 + c^2$, or equivalently $a^2 - b^2 = c^2$. Now clearly $\overline{FG} = \overline{BG} = \overline{HG} = a$, since all are radii of the semicircle. Thus, $\overline{EF} = \overline{FG} - \overline{EG} = a - b$ and $\overline{BE} = \overline{BG} + \overline{GE} = a + b$. It follows that

$$\begin{aligned}\text{Area (rectangle } BCDE) &= (\text{base}) \times (\text{height}) \\ &= (\overline{BE}) \times (\overline{ED}) \\ &= (\overline{BE}) \times (\overline{EF}), \text{ since we constructed } \overline{EF} = \overline{ED} \\ &= (a + b)(a - b) \text{ by the observations above} \\ &= a^2 - b^2 \\ &= c^2 = \text{Area (square } EKLH)\end{aligned}$$

Consequently, we have proved that the original rectangular area equals that of the shaded square which we *constructed* with compass and straightedge, and this completes the rectangle's quadrature.

With this done, the steps toward squaring more irregular regions come quickly.

STEP 2 Quadrature of the triangle (Figure 1.8)

Given $\triangle BCD$, construct a perpendicular from D meeting BC at point E. Of course, we call $\overline{DE}$ the triangle's "altitude" or "height" and know that the area of the triangle is ½(base) × (height) = ½$(\overline{BC}) \times (\overline{DE})$. If we bisect DE at F and construct a rectangle with $\overline{GH} = \overline{BC}$ and $\overline{HJ} = \overline{EF}$, we know that the rectangle's area is $(\overline{HJ}) \times (\overline{GH}) = (\overline{EF}) \times (\overline{BC}) = \frac{1}{2}(\overline{DE}) \times (\overline{BC})$ = area ($\triangle BCD$). But we then apply Step 1 to construct a square equal in area to this rectangle, and so the square's area is also that of $\triangle BCD$. This completes the quadrature of the triangle.

We next move to the following very general situation.

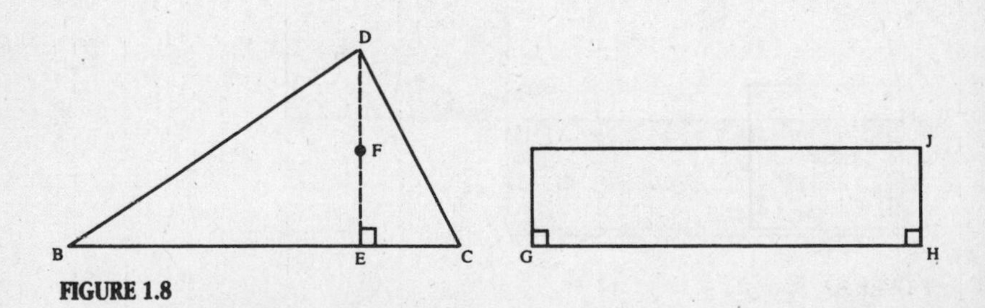

FIGURE 1.8

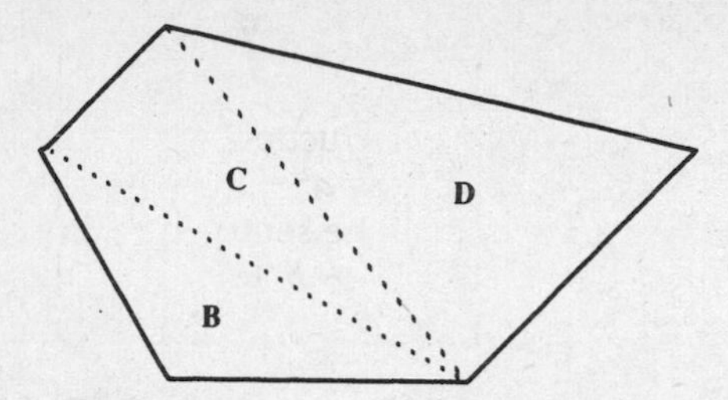

FIGURE 1.9

STEP 3 Quadrature of the polygon (Figure 1.9)

This time we begin with a general polygon, such as the one shown. By drawing diagonals, we subdivide it into a collection of triangles with areas **B**, **C**, and **D**, so that the total polygonal area is **B** + **C** + **D**.

Now triangles are known to be quadrable by Step 2, so we can construct squares with sides b, c, and d and areas **B**, **C**, and **D** (Figure 1.10). We then construct a right triangle with legs of length b and c, whose hypotenuse is of length x, where $x^2 = b^2 + c^2$. Next, we construct a right triangle with legs of length x and d and hypotenuse y, where we have $y^2 = x^2 + d^2$, and finally, the shaded square of side y (Figure 1.11).

Combining our facts, we see that

$$y^2 = x^2 + d^2 = (b^2 + c^2) + d^2 = \mathbf{B} + \mathbf{C} + \mathbf{D}$$

so that the area of the original polygon equals the area of the square having side y.

This procedure clearly could be adapted to the situation in which the polygon was divided by its diagonals into four, five, or any number of triangles. No matter what polygon we are given (see Figure 1.12), we can subdivide it into a set of triangles, square each one by Step 2, and use these individual squares and the Pythagorean theorem to build a

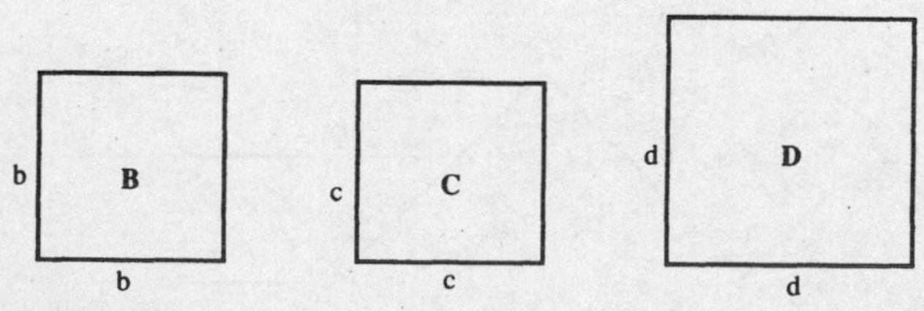

FIGURE 1.10

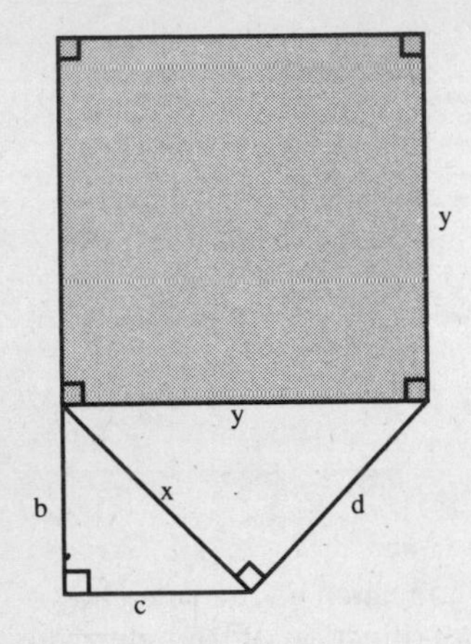

FIGURE 1.11

large square with area equal to that of the polygon. In short, polygons are quadrable.

By an analogous technique we could likewise square a figure whose area was the *difference* between—and not the sum of—two quadrable areas. That is, suppose we knew that area **E** was the difference between areas **F** and **G**, and we had already constructed squares of sides f and g with areas as shown in Figure 1.13. Then we would construct a right triangle with hypotenuse f and leg g. We let e be the length of the other leg and construct a square with side e. We then have

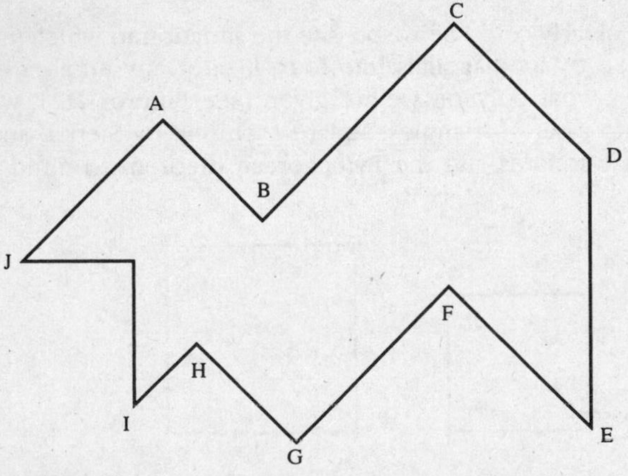

FIGURE 1.12

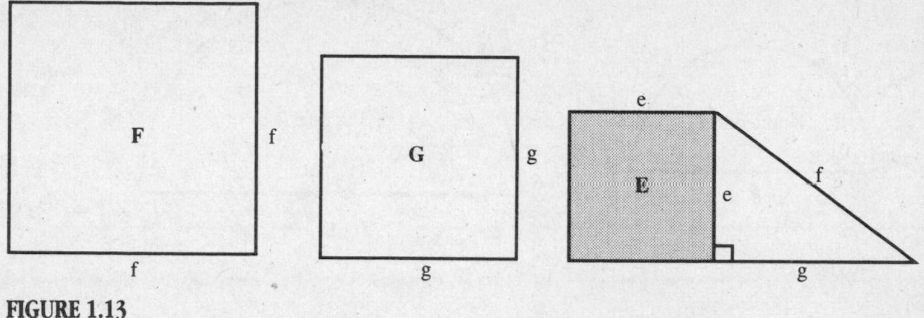

FIGURE 1.13

$$\text{Area (square)} = e^2 = f^2 - g^2 = \mathbf{F} - \mathbf{G} = \mathbf{E}$$

so that area **E** is likewise quadrable.

With the foregoing techniques, the Greeks of Hippocrates' day could square wildly irregular polygons. But this triumph was tempered by the fact that such figures are *rectilinear*—that is, their sides, although numerous and meeting at all sorts of strange angles, are merely straight lines. Far more challenging was the issue of whether figures with curved boundaries—the so-called *curvilinear* figures—were likewise quadrable. Initially, this must have seemed unlikely, for there is no obvious means to straighten out curved lines with compass and straightedge. It must therefore have been quite unexpected when Hippocrates of Chios succeeded in squaring a curvilinear figure known as a "lune" in the fifth century B.C.

Great Theorem: The Quadrature of the Lune

A lune is a plane figure bounded by two circular arcs—that is, a crescent. Hippocrates did not square all such figures but rather a particular lune he had carefully constructed. (As will be shown in the Epilogue, this distinction seemed to be the source of some misunderstanding in later Greek geometry.) His argument rested upon three preliminary results:

- The Pythagorean theorem
- An angle inscribed in a semicircle is right.
- The areas of two circles or semicircles are to each other as the squares on their diameters.

$$\frac{\text{Area (semicircle 1)}}{\text{Area (semicircle 2)}} = \frac{d^2}{D^2}$$

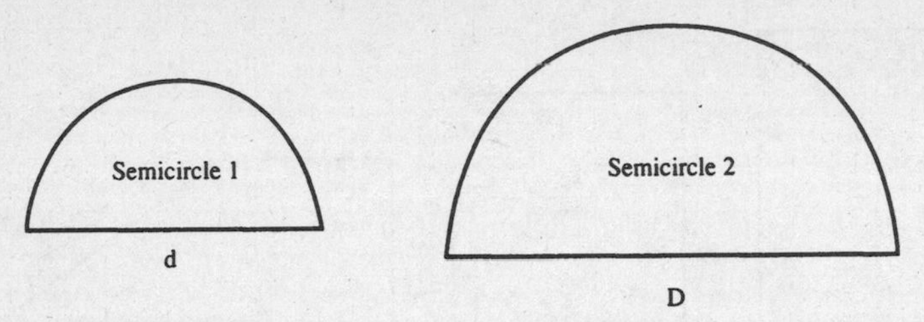

FIGURE 1.14

The first two of these results were well known long before Hippocrates came upon the scene. The last proposition, on the other hand, is considerably more sophisticated. It gives a comparison of the areas of two circles or semicircles based on the relative areas of the squares constructed on their diameters (see Figure 1.14). For instance, if one semicircle has five times the diameter of another, the former has 25 times the area of the latter. This proposition presents math historians with a problem, for there is widespread doubt that Hippocrates actually had a valid proof. He may well have *thought* he could prove it, but modern scholars generally feel that this theorem—which later appeared as the second proposition in Book XII of Euclid's *Elements*—presented logical difficulties far beyond what Hippocrates would have been able to handle. (A derivation of this result is presented in Chapter 4.)

That aside, we now consider Hippocrates' proof. Begin with a semicircle having center O and radius $\overline{AO} = \overline{OB}$, as shown in Figure 1.15. Construct OC perpendicular to AB, with point C on the semicircle, and draw lines AC and BC. Bisect AC at D, and using $\overline{AD}$ as a radius and D as center, draw semicircle AEC, thus creating lune $AECF$, which is shaded in the diagram.

Hippocrates' plan of attack was simple yet brilliant. He first had to establish that the lune in question had *precisely* the same area as the shaded $\triangle AOC$. With this behind him, he could then apply the known fact that triangles can be squared to conclude that the lune can be squared as well. The details of the classic argument follow:

THEOREM Lune *AECF* is quadrable.

PROOF Note that $\angle ACB$ is right since it is inscribed in a semicircle. Triangles AOC and BOC are congruent by the "side-angle-side" congruence scheme, and consequently $\overline{AC} = \overline{BC}$. We thus apply the Pythagorean theorem to get

$$(\overline{AB})^2 = (\overline{AC})^2 + (\overline{BC})^2 = 2(\overline{AC})^2$$

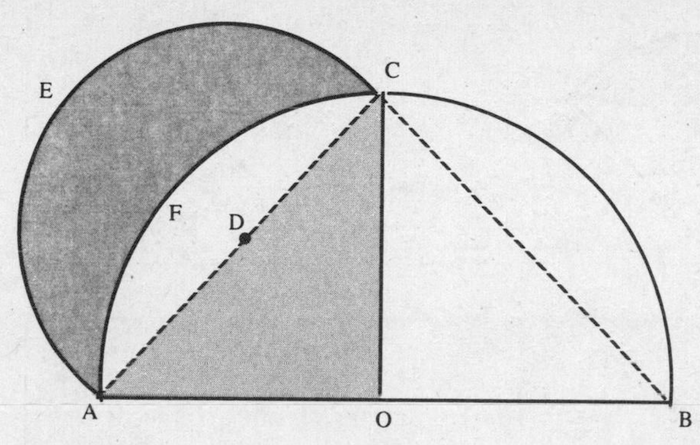

FIGURE 1.15

Because *AB* is the diameter of semicircle *ACB*, and *AC* is the diameter of semicircle *AEC*, we can apply the third principle above to get

$$\frac{\text{Area (semicircle } AEC)}{\text{Area (semicircle } ACB)} = \frac{(\overline{AC})^2}{(\overline{AB})^2} = \frac{(\overline{AC})^2}{2(\overline{AC})^2} = \frac{1}{2}$$

In other words, semicircle *AEC* has half the area of semicircle *ACB*.

But we now look at quadrant *AFCO* (a "quadrant" is a quarter of a circle). Clearly this quadrant also has half the area of semicircle *ACB*, and we immediately conclude that

$$\text{Area (semicircle } AEC) = \text{Area (quadrant } AFCO)$$

Finally, we need only subtract from each of these figures their shared region *AFCD*, as in Figure 1.16. This leaves

$$\begin{aligned}&\text{Area (semicircle } AEC) - \text{Area (region } AFCD)\\&\qquad = \text{Area (quadrant } AFCO) - \text{Area (region } AFCD)\end{aligned}$$

and a quick look at the diagram verifies that this amounts to

$$\text{Area (lune } AECF) = \text{Area } (\triangle ACO)$$

But, as we have seen, we can construct a square whose area equals that of the triangle, and thus equals that of the lune as well. This is the quadrature we sought.

Q.E.D.

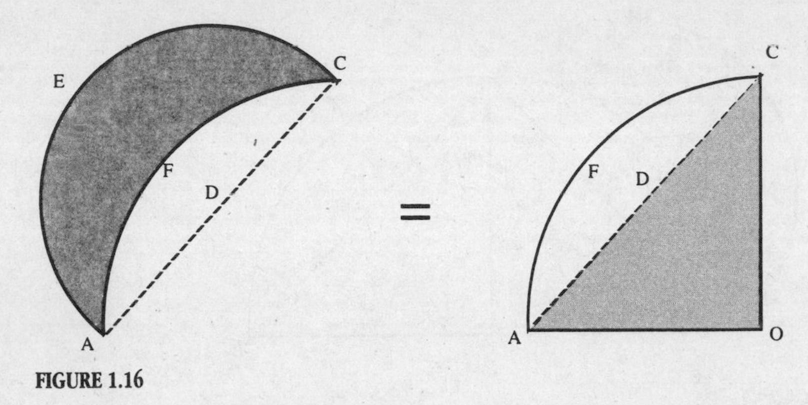

FIGURE 1.16

Here indeed was a mathematical triumph. Looking back from his fifth century vantage point, the commentator Proclus (A.D. 410–485) would write that Hippocrates of Chios ". . . squared the lune and made many other discoveries in geometry, being a man of genius when it came to constructions, if ever there was one."

Epilogue

With Hippocrates' success at squaring the lune, Greek mathematicians must have been optimistic about squaring that most perfect curvilinear figure, the circle. The ancients devoted much time to this problem, and some later writers attributed an attempt to Hippocrates himself, although the matter is again clouded by the difficulties of assessing commentaries upon commentaries. Nonetheless, Simplicius, writing in the fifth century, quoted his predecessor Alexander Aphrodisiensis (ca. A.D. 210) as saying that Hippocrates had claimed that he could square the circle. Piecing together the evidence, we gather that this is the sort of argument Alexander had in mind:

Begin with an arbitrary circle with diameter AB. Construct a large circle with center O and a diameter CD that is *twice* AB. Within the larger circle, inscribe a regular hexagon by the known technique of letting each side be the circle's radius. That is,

$$\overline{CE} = \overline{EF} = \overline{FD} = \overline{DG} = \overline{GH} = \overline{HC} = \overline{OC}$$

It is important to note that each of these segments, being the radius of the larger circle, also has length $\overline{AB}$. Then, using the six segments as diameters, construct the six semicircles shown in Figure 1.17. This gen-

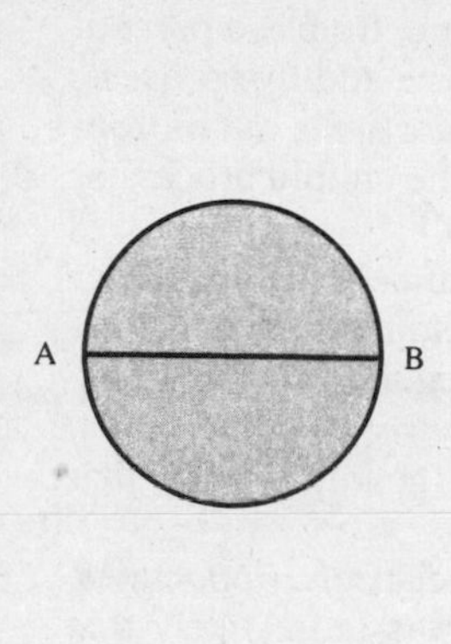

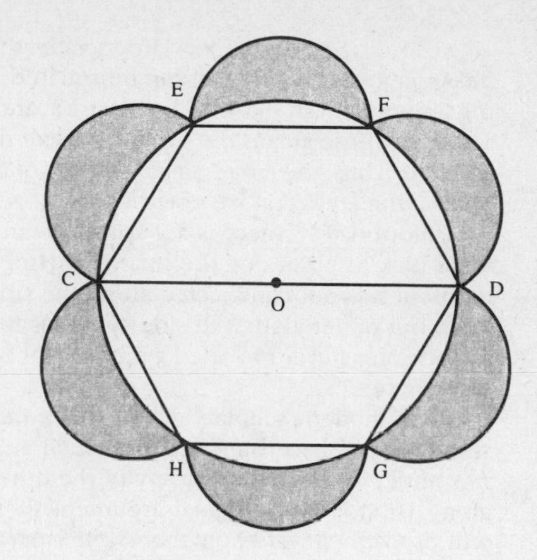

FIGURE 1.17

erates the shaded region composed of the six lunes and the circle upon *AB*.

Next imagine decomposing the figure on the right in two different ways: first, as the regular hexagon *CEFDGH* plus the six semicircles; second, as the large circle plus the six lunes. Obviously these yield the same overall area since they arise from the decomposition of the same figure. But the six semicircles amount to three full circles, each with diameter equal to $\overline{AB}$. Thus,

$$\text{Area (hexagon)} + 3 \text{ Area (circle on } AB) = \text{Area (large circle)} + \text{Area (six lunes)}$$

Now the large circle, having twice the diameter, must have $2^2 = 4$ times the area of its smaller counterpart. Hence,

$$\text{Area (hexagon)} + 3 \text{ Area (circle on } AB) = 4 \text{ Area (circle on } AB) + \text{Area (six lunes)}$$

and, subtracting "3 Area (circle on *AB*)" from both sides of this equation, we get

$$\text{Area (hexagon)} = \text{Area (circle on } AB) + \text{Area (six lunes)} \qquad \text{or}$$

$$\text{Area (circle on } AB) = \text{Area (hexagon)} - \text{Area (six lunes)}$$

According to Alexander, Hippocrates then reasoned as follows: The hexagon, being a polygon, can be squared; each lune, from the preceding argument, can likewise be squared, and so, by the additive process, a square whose area is the sum of the half-dozen lunar areas can be constructed. Thus, the circle on *AB* can be squared by the simple process of subtracting areas that we noted earlier.

Unfortunately, there is a glaring flaw in this argument, as Alexander was quick to point out: the lune that Hippocrates squared in our great theorem was not constructed along the side of a regular inscribed *hexagon* but rather along the side of an inscribed *square*. In other words, Hippocrates never provided a process for squaring the kind of lune that arose here.

Most modern scholars doubt that a mathematician of Hippocrates' stature could have bumbled into such an error. It is more likely that Alexander or Simplicius or any of the other intermediaries who passed along Hippocrates' original argument garbled it in some manner. We will probably never know the whole story. Nonetheless, it is likely that this kind of reasoning supported the idea that the quadrature of the circle should somehow be possible. If the preceding argument did not quite do the job, then maybe just a little more effort and a little more insight might have brought success.

But it was not to be. For generations, for centuries, the challenge to square the circle went unmet, although not for any lack of trying. Countless solutions were proposed involving a multitude of ingenious twists and turns. Yet in the end, each was found to contain an error. Gradually, mathematicians began to suspect that there was an intrinsic impossibility in the circle's quadrature with compass and straightedge. Of course, the mere lack of a correct argument, even after 2000 years of trying, did not establish its impossibility; perhaps mathematicians had just not been clever enough to find their way through the geometric thickets. Further, if the quadrature of the circle was impossible, this fact would have to be *proved* with all the logical rigor of any other theorem, and it was by no means clear how to go about such a proof.

One point should be stressed. No one doubted that, given a circle, there *exists* a square of equal area. For instance, consider a given, fixed circle and a small square spot of light projecting on the page beside it, the square's area being substantially less than that of the circle. If we continuously move the projector away from the page, thereby gradually increasing the area of the square image, we eventually arrive at a square whose area exceeds that of the circle. Appealing to the intuitive notion of "continuous growth," we can correctly conclude that at some intermediate instant, the area of the square exactly equaled the area of the circle.

But this is all beside the point. Remember that the crucial issue is not whether such a square *exists*, but whether it can be *constructed* with compass and straightedge. It is here that the difficulties appeared, for the geometer was limited to these two particular tools; moving spotlights around was simply against the rules.

The problem of squaring the circle remained unresolved from the time of Hippocrates until just over a century ago. At last, in 1882, the German mathematician Ferdinand Lindemann (1852–1939) succeeded in proving unequivocally that the quadrature of the circle was an impossibility. The technical details of his proof are quite advanced and go well beyond the scope of this book. However, the following is a brief synopsis of how it was that Lindemann answered this age-old question.

He did it by translating the issue from the realm of geometry to the realm of number. If we imagine the collection of all real numbers, depicted in the schematic diagram in Figure 1.18 as being contained within the large rectangle, we can subdivide them into two exhaustive and mutually exclusive categories—the algebraic numbers and the transcendental numbers.

By definition, a real number is *algebraic* if it is the solution to some polynomial equation

$$a_n x^n + a_{n-1}x^{n-1} + \ldots + a_2x^2 + a_1x + a_0 = 0$$

where all the coefficients a_n, a_{n-1}, . . . , a_2, a_1, and a_0 are integers. Thus, the rational number $\frac{2}{3}$ is algebraic since it is the solution of the polynomial equation $3x - 2 = 0$; the irrational $\sqrt{2}$ is likewise algebraic since it satisfies $x^2 - 2 = 0$; and even $\sqrt[3]{1 + \sqrt{5}})$ is algebraic since it satisfies $x^6 - 2x^3 - 4 = 0$. Note that, in each case, these polynomials have integer coefficients.

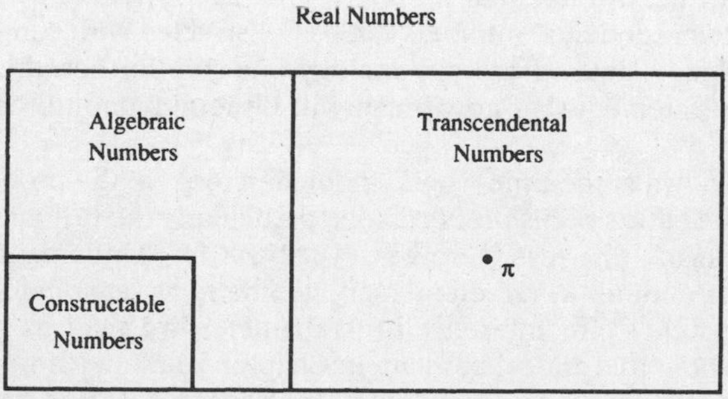

Real Numbers

Algebraic Numbers

Transcendental Numbers

$\bullet\ \pi$

Constructable Numbers

FIGURE 1.18

Less formally, we can think of the algebraic numbers as the "easy" or "familiar" quantities encountered in arithmetic and elementary algebra. For instance, all whole numbers are algebraic, as are all fractions and their square roots, cube roots, and so on.

By contrast, a number is *transcendental* if it is not algebraic—that is, if it is not the solution of *any* polynomial equation with integer coefficients. Such numbers are much more complicated than their relatively simple algebraic cousins. By the very definition, it is clear that any real number is either algebraic or transcendental but not both. This is a stark dichotomy, rather like any person's being either a man or a woman, with no middle ground.

Now begin with a unit length (that is, a length to represent the number "1") and keep track of what other lengths we can produce by straightedge and compass construction. It turns out that the totality of all possible constructible lengths, while vast, does not include every real number. For instance, starting from a length of 1, we can construct lengths of 2, 3, 4, and so on, as well as rational lengths like $1/2$, $2/5$, $13/11$ and even irrational lengths involving only square roots, like $\sqrt{2}$ or $\sqrt{5}$. Further, if we can construct two magnitudes, we can construct their sum, difference, product, or quotient. Putting all of these operations together, we see that more complex expressions such as

$$\sqrt{\frac{6 - 2\sqrt{2}}{1 + \sqrt{4 + \sqrt{23 - \sqrt{7}}}}}$$

are actually constructible lengths.

This vast array of constructible numbers forms a subset of the algebraic numbers, even as the collection of all bald men forms a subset of all men. As Figure 1.18 suggests, these constructible quantities are strictly embedded within the algebraic numbers. The crucial point is that no member of the transcendental numbers can be constructed with compass and straightedge. (If we stretch our analogy one step further, this corresponds to the statement that no woman will be found among the bald men.)

All of this was known at the time when Lindemann took up the problem. Building on the efforts of his predecessors, particularly the brilliant French mathematician Charles Hermite (1822–1901), Lindemann attacked the famous number π. (In elementary geometry we encounter π as the ratio of a circle's circumference to its diameter; we shall have much more to say about this critical constant in Chapter 4.) Lindemann's triumph was to prove that π is transcendental. In other words, π is *not* algebraic and thus is not constructible. This, in turn, tells us that $\sqrt{\pi}$ is not constructible either, since if we could construct $\sqrt{\pi}$, we could, with a few more swipes of the compass and straightedge, construct π as well.

At first, this numerical discovery may seem to have little bearing on the geometry of circle-squaring, but we shall see that it provided the missing piece of the puzzle.

THEOREM The quadrature of the circle is impossible

PROOF Let us assume, for the sake of eventual contradiction, that circles *can* be squared. We get out our compass and easily construct a circle having radius $r = 1$. Its area is thus $\pi r^2 = \pi$. If circles are quadrable, as we have temporarily assumed, then we employ our compass and straightedge, work feverishly slashing arcs and drawing lines, and eventually, after only a finite number of such steps, end up with a square that also has area π, as indicated in Figure 1.19. In this process, we would have had to *construct* the square, which of course would require us to have constructed each of its four sides. Call the length of the square's side x. Then we see that

$$\pi = \text{Area of circle} = \text{Area of square} = x^2$$

and so the length $x = \sqrt{\pi}$ would be constructible with compass and straightedge. But, as we have noted, no such construction for $\sqrt{\pi}$ is possible.

What went wrong? Tracing back through the argument and looking for the source of our contradiction, we find it can only be the initial assumption, namely, that circles can be squared. As a consequence, we must reject this and conclude, once and for all, that the quadrature of the circle is a logical impossibility!

Q.E.D.

Lindemann's discovery, then, showed that squaring the circle—a quest that occupied mathematicians from Hippocrates' day until modern

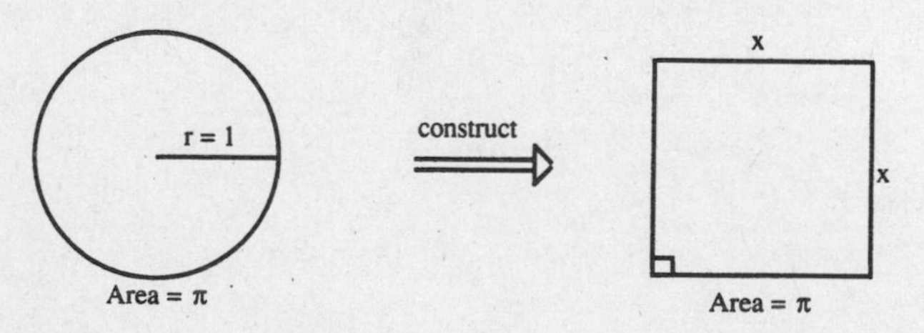

FIGURE 1.19

times—was a lost cause. All of the suggestive proofs, all of the promising clues starting with the quadrature of the lune, turned out to be illusory. Compass and straightedge alone are inadequate for turning circles into squares.

And what did history have to say about lunes? Our great theorem above showed Hippocrates squaring a particular lune, and he managed to do two other kinds as well. Thus, as of 440 B.C., three types of lunes were known to be quadrable. At this point, progress stopped for over two millennia until, in 1771, the great Leonhard Euler (1707–1783)—who will be the object of our attention in Chapters 9 and 10—found two more kinds of lunes that were squarable. There the matter rested until the twentieth century when N. G. Tschebatorew and A. W. Dorodnow proved that these five are the *only* squarable lunes! All other lunes, such as the one that generated Alexander's harsh criticism cited earlier, share with the circle the impossibility of being squared.

So the final chapter in the story of Hippocrates and his lunes has been written, and it has been a rather perverse story at that. At first, intuition suggested that curved figures could not be squared with compass and straightedge. Hippocrates' lunes turned intuition upside down, and the search was on for quadratures galore. But, in the end, the negative results of Lindemann, Tschebatorew, and Dorodnow showed that intuition had not been so flawed after all. The quadrature of curvilinear figures, far from being the norm, must forever remain the exception.

Chapter 2

Euclid's Proof of the Pythagorean Theorem

(ca. 300 B.C.)

The *Elements* of Euclid

A century and a half passed between Hippocrates and Euclid. During this span, Greek civilization grew and matured, enriched by the writings of Plato and Aristotle, of Aristophanes and Thucydides, even as it underwent the turmoil of the Peloponnesian Wars and the glory of the Greek empire under Alexander the Great. By 300 B.C., Greek culture had spread across the Mediterranean world and beyond. In the West, Greece reigned supreme.

The period from 440 B.C. to 300 B.C. saw a number of individuals contribute significantly to the development of mathematics. Among these were Plato (427–347 B.C.) and Eudoxus (ca. 408–355 B.C.), although only the latter was truly a mathematician.

Plato, the great philosopher of Athens, deserves mention here not so much for the mathematics he created as for the enthusiasm and status he imparted to the subject. As a youth, Plato had studied in Athens under Socrates and is of course our primary source of information about his esteemed teacher. For a number of years Plato roamed the world, meeting the great thinkers and formulating his own philosophical positions. In 387 B.C., he returned to his native Athens and founded the Academy. Devoted to learning and contemplation, the Academy attracted talented scholars from near and far, and under Plato's guidance it became the intellectual center of the classical world.

Of the many subjects studied at the Academy, none was more highly regarded than mathematics. The subject certainly appealed to Plato's sense of beauty and order and represented an abstract, ideal world unsullied by the humdrum demands of day-to-day existence. Moreover, Plato considered mathematics to be the perfect training for the mind, its logical rigor demanding the ultimate in concentration, cleverness, and care. Legend has it that across the arched entryway to his prestigious Academy were the words "Let no man ignorant of geometry enter here." Explicit sexism notwithstanding, this motto reflected the view that only those who had first demonstrated a mathematical maturity were capable of facing the intellectual challenge of the Academy. We might say that Plato regarded geometry as the ideal entrance requirement, the Scholastic Aptitude Test of his day.

Although very little original mathematics is now attributed to Plato, the Academy produced many capable mathematicians and one indisputably great one, Eudoxus of Cnidos. Eudoxus came to Athens about the time the Academy was being created and attended the lectures of Plato himself. Eudoxus' poverty forced him to live in Piraeus, on the outskirts of Athens, and make the daily round-trip journey to and from the Academy, thus distinguishing him as one of the first commuters (although we are unsure whether he had to pay out-of-city-state tuition). Later in his career, he traveled to Egypt and returned to his native Cnidos, all the while assimilating the discoveries of science and constantly extending its frontiers. Particularly interested in astronomy, Eudoxus devised complex explanations of lunar and planetary motion whose influence was felt until the Copernican revolution in the sixteenth century. Never willing to accept divine or mystical explanations for natural phenomena, he instead tried to subject them to observation and rational analysis. Thus, Sir Thomas Heath said of Eudoxus, "He was a *man of science* if ever there was one."

In mathematics, Eudoxus is remembered for two major contributions. One was his theory of proportion, and the other his method of exhaustion. The former provided a logical victory over the impasse created by the Pythagoreans' discovery of incommensurable magnitudes. This impasse was especially apparent in geometric theorems about similar triangles, theorems that had initially been proved under the assumption that *any* two magnitudes were commensurable. When this assumption was destroyed, so too were the existing proofs of some of geometry's foremost theorems. What resulted is sometimes called the "logical scandal" of Greek geometry. That is, while people still believed that the theorems were correct as stated, they no longer were in possession of sound proofs with which to support this belief. It was Eudoxus who developed a valid theory of proportions and thereby supplied the long-sought proofs. His theory, which must have brought a collective sigh of relief from the Greek mathematical world, is now most readily found in Book V of Euclid's *Elements.*

Eudoxus' other great contribution, the method of exhaustion, found immediate application in the determination of areas and volumes of the more sophisticated geometric figures. The general strategy was to approach an irregular figure by means of a succession of known elementary ones, each providing a better approximation than its predecessor. We can think, for instance, of a circle as being a totally curvilinear, and thus quite intractable, plane figure. But, if we inscribe within it a square, and then double the number of sides of the square to get an octagon, and then again double the number of sides to get a 16-gon, and so on, we will find these relatively simple polygons ever more closely approximating the circle itself. In Eudoxean terms, the polygons are "exhausting" the circle from within.

This process is, in fact, precisely how Archimedes determined the area of a circle, as we shall see in the great theorem of Chapter 4. It is to Eudoxus that he owed this fundamental logical tool. In addition, Archimedes credited Eudoxus with using the method of exhaustion to prove that the volume of "any cone is one third part of the cylinder which has the same base with the cone and equal height," a theorem that is by no means trivial. The reader familiar with higher mathematics will recognize in the method of exhaustion the geometric forerunner of the modern notion of "limit," which in turn lies at the heart of the calculus. Eudoxus' contribution was a significant one, and he is usually regarded as being the finest mathematician of antiquity next to the unsurpassed Archimedes himself.

It was during the latter third of the fourth century B.C. that Alexander the Great emerged from Macedonia and set out to conquer the world. His conquests carried him to Egypt where, in 332 B.C., he established the city of Alexandria at the mouth of the Nile River. This city grew rapidly, reportedly reaching a population of half a million in the next three decades. Of particular importance was the formation of the great Alexandrian Library that soon supplanted the Academy as the world's foremost center of scholarship. At one point, the facility had over 600,000 papyrus rolls, a collection far more complete and astounding than anything the world had ever seen. Indeed, Alexandria would remain the intellectual focus of the Mediterranean world through the Greek and Roman periods until its final destruction in A.D. 641 at the hands of the Arabs.

Among the scholars attracted to Alexandria around 300 B.C. was a man named Euclid, who came to set up a school of mathematics. We know very little about his life either before or after his arrival on the African coast, but it appears that he received his training at the Academy from the followers of Plato. Be that as it may, Euclid's influence was so profound that virtually all subsequent Greek mathematicians had some connection or other with the Alexandrian School.

What Euclid did that established him as one of the greatest names in mathematics history was to write the *Elements*. This work had a profound impact on Western thought as it was studied, analyzed, and edited for century upon century, down to modern times. It has been said that of all books from Western civilization, only the Bible has received more intense scrutiny than Euclid's *Elements*.

The highly acclaimed *Elements* was simply a huge collection—divided into 13 books—of 465 propositions from plane and solid geometry and from number theory. Today, it is generally agreed that relatively few of these theorems were of Euclid's own invention. Rather, from the known body of Greek mathematics, he created a superbly organized treatise that was so successful and so revered that it thoroughly obliterated all preceding works of its kind. Euclid's text soon became the standard. Consequently, a mathematician's reference to I.47 can only mean the 47th proposition of the first book of the *Elements;* there is no more need to *say* that we are talking about the *Elements* than there is to specify that I Kings 7:23 is referring to the Bible.

Actually the parallel is quite accurate, for no book has come closer to being the "bible of mathematics" than Euclid's spectacular creation. Down through the centuries, over 2000 editions of the *Elements* have appeared, a figure that must make the authors of today's mathematics textbooks drool with envy. As noted, it was highly successful even in its own day. After the fall of Rome, the Arab scholars carried it off to Baghdad, and when it reappeared in Europe during the Renaissance, its impact was profound. The work was studied by the great Italian scholars of the sixteenth century and by a young Cambridge student named Isaac Newton a century later. We have a passage from Carl Sandburg's biography of Abraham Lincoln that recounts how, when a young lawyer trying to sharpen his reasoning skills, the largely unschooled Lincoln

> . . . bought the *Elements* of Euclid, a book twenty-three centuries old . . . [It] went into his carpetbag as he went out on the circuit. At night . . . he read Euclid by the light of a candle after others had dropped off to sleep.

It has often been noted that Lincoln's prose was influenced and enriched by his study of Shakespeare and the Bible. It is likewise obvious that many of his political arguments echo the logical development of a Euclidean proposition.

And Bertrand Russell (1872–1970) had his own fond memories of the *Elements*. In his autobiography, Russell penned this remarkable recollection:

> At the age of eleven, I began Euclid, with my brother as tutor. This was one of the great events of my life, as dazzling as first love.

As we consider the *Elements* in this chapter and the next, we should be aware that we proceed along paths that so many others have trod. Only a very few classics—the *Iliad* and *Odyssey* come to mind—share such a heritage. The propositions we shall examine were studied by Archimedes and Cicero, by Newton and Leibniz, by Napoleon and Lincoln. It is a bit daunting to place oneself in this long, long line of students.

Euclid's great genius was not so much in creating a new mathematics as in presenting the old mathematics in a thoroughly clear, organized, and logical fashion. This is no small accomplishment. It is important to recognize the *Elements* as more than just mathematical theorems and their proofs; after all, mathematicians as far back as Thales had been furnishing proofs of propositions. Euclid gave us a splendid *axiomatic* development of his subject, and this is a critical distinction. He began the *Elements* with a few basics: 23 definitions, 5 postulates, and 5 common notions or general axioms. These were the foundations, the "givens," of his system. He could use them at any time he chose. From these basics, he proved his first proposition. With this behind him, he could then blend his definitions, postulates, common notions, and this first proposition into a proof of his second. And on it went.

Consequently, Euclid did not just furnish proofs; he furnished them within this axiomatic framework. The advantages of such a development are significant. For one thing, it avoids circularity in reasoning. Each proposition has a clear, unambiguous string of predecessors leading back to the original axioms. Those familiar with computers could even draw a flow chart showing precisely which results went into the proof of a given theorem. This approach is far superior to "plunging in" to prove a proposition, for in such a case it is never clear which previous results can and cannot be used. The great danger from starting in the middle, as it were, is that to prove theorem A, one might need to use result B, which, it may turn out, cannot be proved without using theorem A itself. This results in a circular argument, the logical equivalent of a snake swallowing its own tail; in mathematics it surely leads to no good.

But the axiomatic approach has another benefit. Since we can clearly pick out the predecessors of any proposition, we have an immediate sense of what happens if we should alter or eliminate one of our basic postulates. If, for instance, we have proved theorem A without ever using either postulate C or any result previously proved by means of postulate C, then we are assured that our theorem A remains valid even if postulate C is discarded. While this might seem a bit esoteric, just such an issue arose with respect to Euclid's controversial fifth postulate and led to one of the longest and most profound debates in the history of mathematics. This matter is examined in the Epilogue of the current chapter.

Thus, the axiomatic development of the *Elements* was of major importance. Even though Euclid did not quite pull this off flawlessly, the high level of logical sophistication and his obvious success at weaving the pieces of his mathematics into a continuous fabric from the basic assumptions to the most sophisticated conclusions served as a model for all subsequent mathematical work. To this day, in the arcane fields of topology or abstract algebra or functional analysis, mathematicians will first present the axioms and then proceed, step-by-step, to build up their wonderful theories. It is the echo of Euclid, 23 centuries after he lived.

Book I: Preliminaries

In this chapter, we shall focus only on the first book of the *Elements*; subsequent books will be the topic of Chapter 3. Book I began abruptly with a list of definitions from plane geometry. (All Euclidean quotations are taken from Sir Thomas Heath's encyclopedic edition *The Thirteen Books of Euclid's Elements.*) Among the first few definitions were:

- **Definition 1** A *point* is that which has no part.
- **Definition 2** A *line* is breadthless length.
- **Definition 4** A *straight line* is a line which lies evenly with the points on itself.

Today's students of Euclid find these statements unacceptable and a bit quaint. Obviously, in any logical system, not every term can be defined, since definitions themselves are composed of terms, which in turn must be defined. If a mathematician tries to give a definition for *everything*, he or she is condemned to a huge circular jumble. What, for instance, did Euclid mean by "breadthless"? What is the technical meaning of lying "evenly with the points on itself"?

From a modern viewpoint, a logical system begins with a few undefined terms to which all subsequent definitions relate. One surely tries to keep the number of these undefined terms to a minimum, but their presence is unavoidable. For modern geometers, then, the notions of "point" and "straight line" remain undefined. Statements such as

Euclid's may serve to convey some image in our minds, and this is not without merit; but as precise, logical definitions, these first few items are unsatisfactory.

Fortunately, his later definitions were more successful. A few of these figure prominently in our discussion of Book I and deserve comment.

□ **Definition 10** When a straight line standing on another straight line makes the adjacent angles equal to one another, each of the equal angles is *right* and the straight line standing on the other is called a *perpendicular* to that on which it stands.

It may come as a surprise to modern readers that Euclid did not define a right angle in terms of 90°; in fact, nowhere in the *Elements* is "degree" ever mentioned as a unit of angular measure. The only angular measure that plays any significant role in the book is the right angle, and as we can see, Euclid defined this as one of two equal adjacent angles along a straight line.

□ **Definition 15** A *circle* is a plane figure contained by one line such that all the straight lines falling upon it from one point among those lying within the figure are equal to one another.

Clearly, the "one point" within the circle is the circle's center, and the equal "straight lines" he referred to are the radii.

In definitions 19 through 22, Euclid defined *triangles* (plane figures contained by three straight lines), *quadrilaterals* (those contained by four), and such specific subclasses as *equilateral* triangles (triangles with three sides equal) and *isosceles* triangles (those with "two of its sides alone equal"). His final definition proved to be critical:

□ **Definition 23** *Parallel* straight lines are straight lines which, being in the same plane and being produced indefinitely in both directions, do not meet one another in either direction.

Notice that Euclid avoided defining parallels in terms of their being everywhere equidistant. His definition was far simpler and less fraught with logical pitfalls: parallels were simply lines in the same plane that never intersect.

With the definitions behind him, Euclid gave a list of five postulates for his geometry. Recall, these were to be the givens, the self-evident truths of his system. He certainly had to select them judiciously and to avoid overlap or internal inconsistency.

POSTULATE 1 [It is possible] to draw a straight line from any point to any point.

POSTULATE 2 [It is possible] to produce a finite straight line continuously in a straight line.

A moment's thought shows that the first two postulates permitted precisely the sorts of constructions one can make with an unmarked straightedge. For instance, if the geometer wanted to connect two points with a straight line—a task physically accomplished with a straightedge—then Postulate 1 provided the logical justification for doing so.

POSTULATE 3 [It is possible] to describe a circle with any center and distance (i.e., radius).

Here was the corresponding logical basis for pulling out a compass and drawing a circle, provided one first had a given point to be the center and a given distance to serve as radius. Thus, the first three postulates, together, justified all pertinent uses of the Euclidean tools.

Or did they? Those who think back to their own geometry training will recall an additional use of the compass, namely, as a means of transferring a fixed length from one part of the plane to another. That is, given a line segment whose length was to be copied elsewhere, one puts the point of the compass at one end of the segment and the pencil tip at the other; then, holding the device rigidly, we lift the compass and carry it to the desired spot. It is a simple and highly useful procedure. However, in playing by Euclid's rules, it was not permitted, for nowhere did he give a postulate allowing this kind of transfer of length. As a result, mathematicians often refer to the Euclidean compass as "collapsible." That is, although it is perfectly capable of drawing circles (as Postulate 3 guaranteed), upon lifting it from the plane, it falls shut, unable to remain open once it is removed.

What is one to make of this situation? Why did Euclid not insert an additional postulate to support this very important transfer of lengths? The answer is simple: he did not need to *assume* such a technique as a postulate, for he *proved* it as the third proposition of Book I. That is, Euclid introduced a clever technique for transferring lengths even if his compass "collapsed" upon lifting it from the page, and then he proved why his technique worked. It is to Euclid's great credit that he avoided assuming what he could in fact derive, and thereby kept his postulates to a bare minimum.

POSTULATE 4 All right angles are equal to one another.

This postulate did not relate to a construction. Rather, it provided a uniform standard of comparison throughout Euclid's geometry. Right angles had been introduced in Definition 10, and now Euclid was assuming that any two such angles, regardless of where they were situated in the plane, were equal. With this behind him, Euclid arrived at by far the most controversial statement in Greek mathematics:

POSTULATE 5 If a straight line falling on two straight lines make the interior angles on the same side less than two right angles, the two straight lines, if produced indefinitely, meet on that side on which are the angles less than the two right angles.

As shown in Figure 2.1, this postulate is saying that if $\alpha + \beta < 2$ right angles, then lines *AB* and *CD* meet toward the right. Postulate 5 is often called Euclid's parallel postulate. This is a bit of a misnomer, since actually the postulate gave conditions under which two lines meet and thus, according to Definition 23, is more accurately called the nonparallel postulate.

Clearly, this postulate was quite unlike the others. It was longer to state, required a diagram to understand, and seemed far from being a self-evident truth. The postulate appeared too complicated to be included in the same category as the innocuous "All right angles are

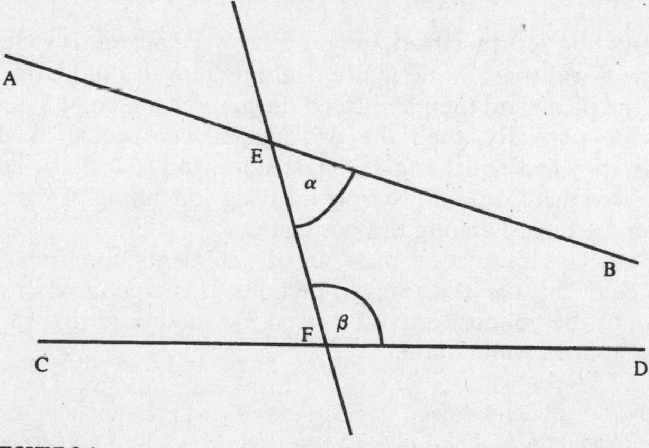

FIGURE 2.1

equal." In fact, many mathematicians felt in their bones that the fifth postulate was, in reality, a theorem. They sensed that, just as Euclid did not need to assume that lengths could be transferred with a compass, neither did he have to assume this postulate; he should simply have been able to prove it from the more elementary properties of geometry. There is evidence that Euclid himself was a bit uneasy about this matter, for in his development of Book I he avoided using the parallel postulate as long as he could. That is, whereas he felt perfectly content to use any of his other postulates as early and often as he needed, Euclid put off the use of his fifth postulate through his first 28 propositions. As shown in the Epilogue, however, it was one thing to be skeptical of the need for such a postulate but quite another to furnish the actual proof.

With this controversial statement behind him, Euclid completed his preliminaries with a list of five common notions. These too were meant to be self-evident truths but were of a more general nature, not specific to geometry. They were

- **Common Notion 1** Things which are equal to the same thing are also equal to one another.
- **Common Notion 2** If equals be added to equals, the wholes are equal.
- **Common Notion 3** If equals be subtracted from equals, the remainders are equal.
- **Common Notion 4** Things which coincide with one another are equal to one another.
- **Common Notion 5** The whole is greater than the part.

Of these, only the fourth raised some eyebrows. Apparently, what Euclid meant by it was that, if one figure could be moved rigidly from one portion of the plane and then be placed down upon a second figure so as to coincide perfectly, then the two figures were equal in all aspects—that is, they had equal angles, equal sides, and so forth. It has long been observed that Common Notion 4, having something of a geometric character, belonged among the postulates.

This, then, was the foundation of assumed statements upon which the entire edifice of the *Elements* was to be built. It is a good point at which to return to the young Bertrand Russell for another of his wonderful autobiographical confessions:

> I had been told that Euclid proved things, and was much disappointed that he started with axioms. At first, I refused to accept them unless my brother could offer me some reason for doing so, but he said, "If you don't accept

them, we cannot go on," and, as I wished to go on, I reluctantly admitted them.

Book I : The Early Propositions

With the preliminaries behind him, Euclid was ready to prove the first of 48 propositions in Book I. Only those propositions of particular interest or importance are discussed here, the goal being to arrive at Propositions I.47 and I.48, which stand as the logical climax of the first book.

If someone were about to develop geometry from a few selected axioms, what would be his or her very first proposition? For Euclid, it was

PROPOSITION 1.1 On a given finite straight line, to construct an equilateral triangle.

PROOF Euclid began with the given segment *AB*, as shown in Figure 2.2. Using *A* as center and *AB* as radius, he constructed a circle; then, with *B* as center and *AB* again as radius, he constructed a second circle. Both constructions, of course, made use of Postulate 3, and neither required the compass to remain open when lifted from the page. Letting *C* be the point where the circles intersect, Euclid invoked Postulate 1 to draw lines *CA* and *CB* and then claimed that $\triangle ABC$ was equilateral. For, by Definition 15, $\overline{AC} = \overline{AB}$ and $\overline{BC} = \overline{AB}$ since these are radii of their respective circles. Then, since Common Notion 1 states that things equal to the same thing are themselves equal, we conclude that $\overline{AC} = \overline{AB} = \overline{BC}$ and so the triangle is equilateral by definition.

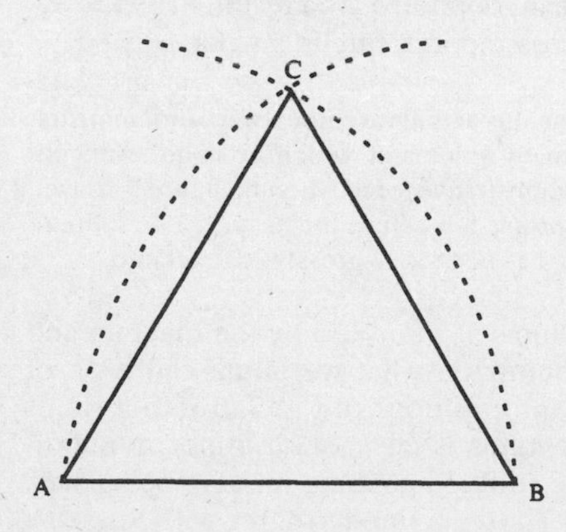

FIGURE 2.2

This was a very simple proof, using two postulates, one common notion, and two definitions, and at first glance it appears perfectly satisfactory. Unfortunately, the proof is flawed. Even the ancient Greeks, no matter how highly they regarded the *Elements*, were aware of the logical shortcomings of this first Euclidean argument.

The problem resided in the point *C*, for how could Euclid prove that the two circles did, in fact, intersect at all? How did he know that they did not somehow pass through one another without meeting? Clearly, since this was his first proposition, he had not previously proved that they must meet. Moreover, nothing in his postulates or common notions spoke to this matter. The only justification for the existence of the point *C* was that it showed up plainly in the diagram.

But this was the rub. For if there was one thing that Euclid wanted to banish from his geometry, it was the reliance on pictures to serve as proofs. By his own ground rules, the proof must rest upon the logic, upon the careful development of the theory from the postulates and common notions, with all conclusions ultimately dependent upon them. When he "let the picture do the talking," Euclid violated the very rules he had imposed upon himself. After all, if we are willing to draw conclusions from the diagrams, we could just as well prove Proposition 1.1 by observing that the triangle thus constructed *looks* equilateral. When we resort to such visual judgments, all is lost.

Modern geometers have recognized the need for an additional postulate, sometimes called the "postulate of continuity," as a justification for claiming that the circles do meet. Other postulates have been introduced to fill similar gaps appearing here and there in the *Elements*. Around the turn of the present century, the mathematician David Hilbert (1862–1943) developed his geometry from a list of 20 postulates, thereby plugging the many Euclidean loopholes. As a result, in 1902 Bertrand Russell gave this negative assessment of Euclid's work:

> His definitions do not always define, his axioms are not always indemonstrable, his demonstrations require many axioms of which he is quite unconscious. A valid proof retains its demonstrative force when no figure is drawn, but very many of Euclid's earlier proofs fail before this test . . . The value of his work as a masterpiece of logic has been very grossly exaggerated.

Admittedly, when he allowed himself to be led by the diagram and not the logic behind it, Euclid committed what we might call a sin of omission. Yet nowhere in all 465 propositions did he fall into a sin of commission. None of his 465 theorems is false. With minor modifications in some of his proofs and the addition of some missing postulates, all have withstood the test of time. Those inclined to agree with Russell's indictment might first compare Euclid's record with that of Greek astronomers or chemists or physicians. These scientists were truly primitive by modern standards, and no one today would rely on ancient texts to explain the motion of the moon or the workings of the liver. But, more often than not, we can rely on Euclid. His work stands as a remarkably timeless achievement. It did not, after all, depend on the collection of data or the creation of more accurate instruments. It rested squarely upon the keenness of reason, and of this Euclid had an abundance.

Propositions I.2 and I.3 cleverly established the previously mentioned ability to transfer a length without an explicit postulate for moving a rigid compass, while Proposition 1.4 was the first of Euclid's congruence schemes. In modern terms, this was the so-called side-angle-side or SAS congruence pattern, which readers should recall from their high school geometry courses. It said that if two triangles have the two sides and included angle of one respectively equal to two sides and included angle of the other, then the triangles are congruent in all respects (Figure 2.3).

In his proof, Euclid assumed that $\overline{AB} = \overline{DE}$, that $\overline{AC} = \overline{DF}$, and that $\angle BAC = \angle EDF$. Then, picking up $\triangle DEF$ and moving it over onto $\triangle ABC$, he argued that the triangles coincided in their entirety. Such a proof by superposition has long since gone out of favor. After all, who is to say that, as figures move around the plane, they are not somehow deformed or distorted? Recognizing the danger here, Hilbert essentially made SAS his Axiom IV.6.

Proposition I.5 stated that the base angles of an isosceles triangle are equal. This theorem came to be known as the "Pons Asinorum," or the bridge of fools. The name stemmed in part from Euclid's diagram, which vaguely resembled a bridge, and in part from the fact that many weaker geometry students could not follow its logic and thus could not cross over into the rest of the *Elements*.

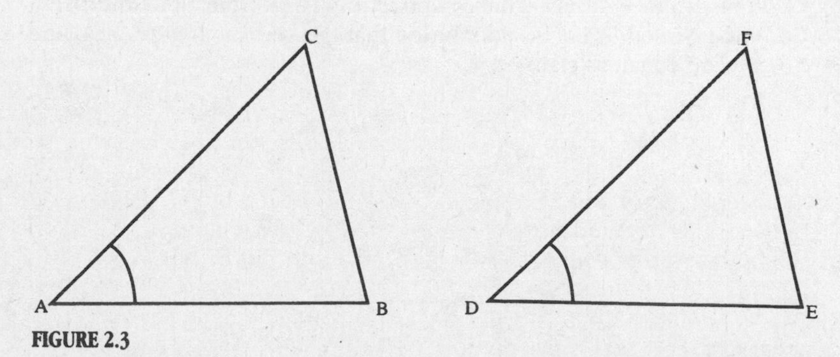

FIGURE 2.3

The following proposition, I.6, was the converse of I.5 in that it stated that if a triangle has base angles equal, then the triangle is isosceles. Of course, theorems and their converses are of great interest to logicians, and often after Euclid had proved a proposition, he would insert the proof of the converse even if it could have been omitted or delayed without otherwise damaging the logical flow of his work.

Euclid's second congruence scheme for triangles—side-side-side or SSS—appeared as Proposition I.8. It stated that when two triangles have the three sides of one respectively equal to the three sides of the other, then their corresponding angles are likewise equal.

Some constructions followed. Euclid demonstrated how to bisect a given angle (Proposition I.9) and a given segment (Proposition 1.10) with compass and straightedge. The subsequent two results showed how to construct a perpendicular to a given line, where the perpendicular either met the line at a given point on it (Proposition I.11) or was drawn downward from a point not on it (Proposition I.12).

Euclid's next two theorems concerned the adjacent angles $\angle ABC$ and $\angle ABD$, as shown in Figure 2.4. In Proposition I.13, he proved that if *CBD* is a straight line, then these two angles sum to two right angles; in I.14, he proved the converse, namely, if $\angle ABC$ and $\angle ABD$ sum to two right angles, then *CBD* is straight. He used this property of angles around a straight line in the simple yet important Proposition I.15.

PROPOSITION I.15 If straight lines cut one another, they make the vertical angles equal to one another (Figure 2.5).

PROOF Since *AEB* is a straight line, Proposition I.13 guaranteed that $\angle AEC$ and $\angle CEB$ sum to two right angles. The same can be said for $\angle CEB$ and $\angle BED$. Now, Postulate 4 stated that all right angles were equal, and this, along with Common Notions 1 and 2 yielded that $\angle AEC + \angle CEB = \angle CEB + \angle BED$. Then, subtracting $\angle CEB$ from both and using Common Notion 3, Euclid concluded that the vertical angles $\angle AEC$ and $\angle BED$ were equal, as claimed.

Q.E.D.

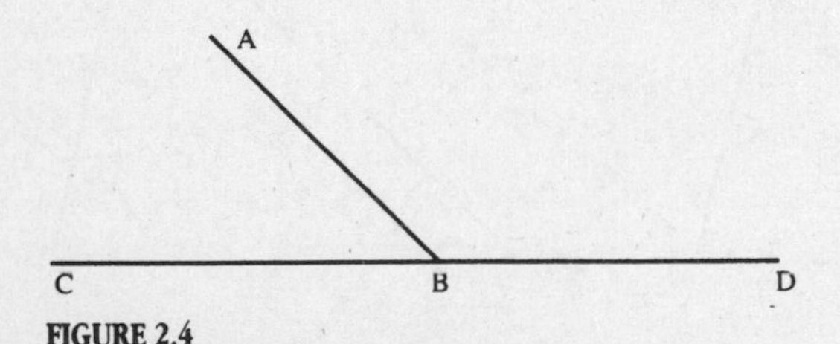

FIGURE 2.4

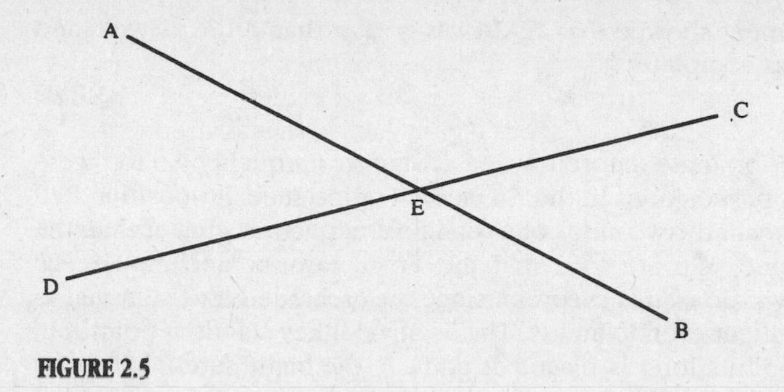

FIGURE 2.5

This brings us to Proposition I.16, the so-called exterior angle theorem, and one of the most important in Book I.

PROPOSITION I.16 In any triangle, if one of the sides be produced, the exterior angle is greater than either of the interior and opposite angles.

PROOF Given $\triangle ABC$ with BC extended to D, as shown in Figure 2.6, we must prove that $\angle DCA$ is greater than either $\angle CBA$ or $\angle CAB$. To begin, Euclid bisected AC at E, by I.10, and then drew line BE by his first postulate. Postulate 2 allowed him to extend BE and he then constructed $\overline{EF} = \overline{EB}$ by I.3. His final construction was to draw FC.

Looking at triangles AEB and CEF, Euclid noted that $\overline{AE} = \overline{CE}$ by the bisection; that vertical angles $\angle 1$ and $\angle 2$ are equal by I.15; and that $\overline{EB} = \overline{EF}$ by construction. Thus, the two triangles are congruent by I.4 (i.e., by SAS), and it follows that $\angle BAE$ equals $\angle FCE$. But $\angle DCA$ is clearly greater than $\angle FCE$, since, by Common Notion 5, the whole is greater than the part. Consequently, exterior $\angle DCA$ exceeds opposite, interior $\angle BAC$. A

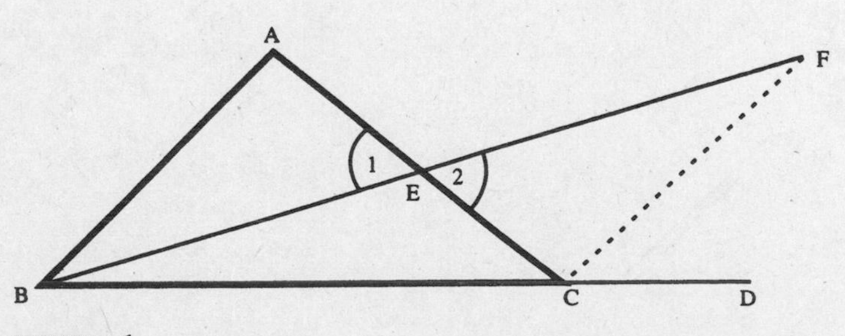

FIGURE 2.6

similar argument showed that $\angle DCA$ was greater than $\angle ABC$ as well, and the proof was complete.

Q.E.D.

The exterior angle theorem was a geometric inequality. So too were the next few propositions in the *Elements.* For instance, Proposition I.20 established that any two sides of a triangle are together greater than the remaining one. We are told that the Epicureans of ancient Greece thought very little of this theorem, since they regarded it as so trivial as to be self-evident even to an ass. That is, if a donkey stands at point *A* in Figure 2.7 and its food is placed at point *B*, the beast surely knows by instinct that the direct route from *A* to *B* is shorter than going along the two sides from *A* to *C* and then from *C* to *B*. It has been suggested that Proposition I.20 is really a self-evident truth and thus should be included among the postulates. However, as with the collapsible compass, Euclid certainly did not want to assume a statement as a postulate if he could prove it as a proposition, and the actual proof he furnished for this theorem was quite a nice bit of logic.

A few more inequality propositions followed before Euclid arrived at the important I.26, his final congruence theorem. Here he first gave a proof of the angle-side-angle, or ASA, congruence scheme as a consequence of the SAS congruence of I.4. But, as the second part of I.26, Euclid gave a fourth, and final, congruence pattern, namely the "angle-angle-side" scheme. That is, he proved that, if $\angle 2 = \angle 5$, $\angle 3 = \angle 6$, and $\overline{AB} = \overline{DE}$ in Figure 2.8, then the triangles *ABC* and *DEF* are themselves congruent.

At first, one is tempted to dismiss this as an immediate consequence of ASA. That is, we can easily see that $\angle 2 + \angle 3 = \angle 5 + \angle 6$, and we could argue that

$$\angle 1 = 2 \text{ right angles} - (\angle 2 + \angle 3) = 2 \text{ right angles} - (\angle 5 + \angle 6) = \angle 4$$

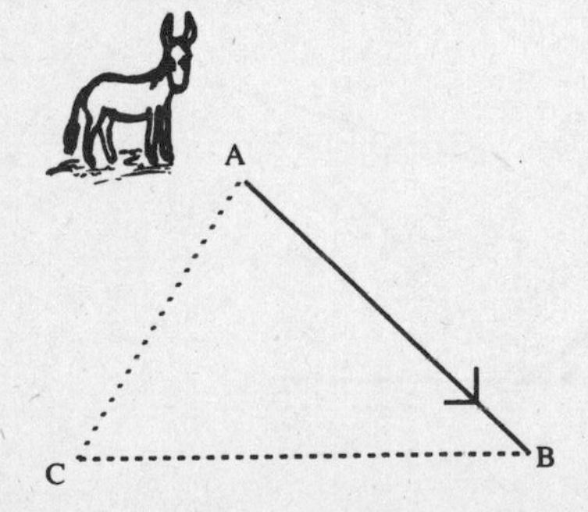

FIGURE 2.7

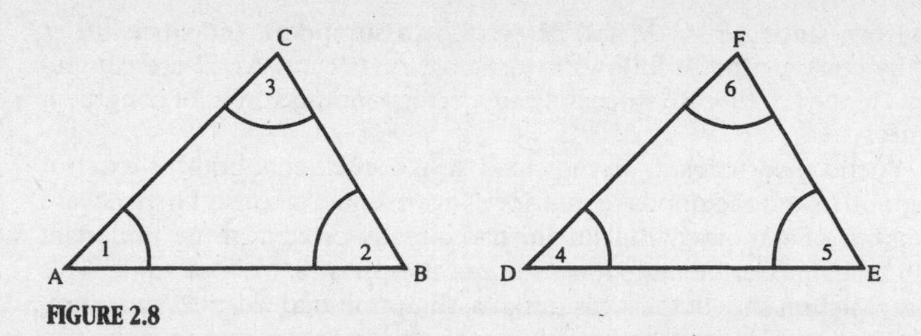

FIGURE 2.8

The congruence would then revert to the ASA scheme, since we have established equality of the angles at either end of the equal sides *AB* and *DE*.

This is a short proof; unfortunately, it is also fallacious. Euclid could not have inserted it at this juncture because he had yet to prove that the sum of the three angles of a triangle totals two right angles. Indeed, without this key theorem it might seem impossible to prove the AAS scheme at all. But Euclid did, with the following elegant proof by contradiction.

PROPOSITION I.26 (AAS) If two triangles have the two angles equal to two angles respectively, and one side equal to one side, namely, . . . that subtending one of the equal angles, they will also have the remaining sides equal to the remaining sides and the remaining angle equal to the remaining angle.

PROOF Consider Figure 2.9. By hypothesis $\angle 2 = \angle 5$, $\angle 3 = \angle 6$, and $\overline{AB} = \overline{DE}$. Euclid claimed that sides *BC* and *EF* must then be equal as well. To prove this he assumed, on the contrary, that one side was longer than the other; for instance, suppose $\overline{BC} > \overline{EF}$. It was thus possible to construct segment *BH* equal in length to *EF*. Draw segment *AH*.

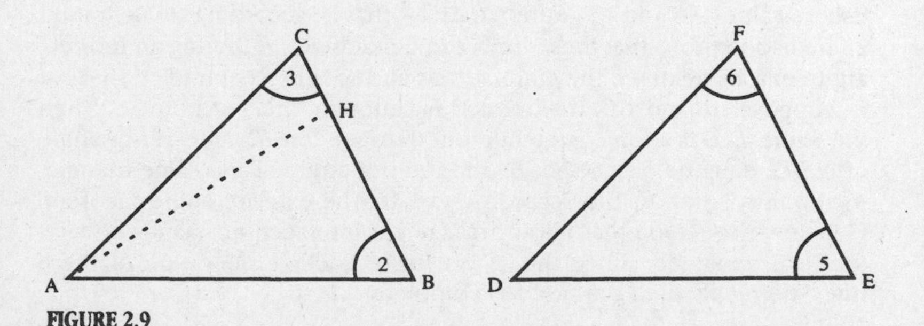

FIGURE 2.9

Now, since $\overline{AB} = \overline{DE}$ and $\angle 2 = \angle 5$ by assumption, and since $\overline{BH} = \overline{EF}$ by construction; it follows by SAS that $\triangle ABH$ and $\triangle DEF$ are congruent. Hence $\angle AHB = \angle 6$ since they are corresponding angles of congruent figures.

Euclid then focused on the small $\triangle AHC$. Note that both its exterior angle *AHB* and the opposite interior $\angle 3$ were equal to $\angle 6$ and hence were equal to one another. But Euclid had already proved in the important I.16 that an exterior angle must *exceed* an opposite, interior angle. This contradiction showed that his initial assumption that $\overline{BC} \neq \overline{EF}$ was invalid. He concluded that these sides were, in fact, equal and thus the two original triangles *ABC* and *DEF* are congruent by SAS.

Q.E.D

Again, note the significance of this clever argument: the four congruence patterns SAS, SSS, ASA, and AAS all hold *without* any reference to the angles of a triangle summing to two right angles.

Proposition I.26 concluded the first part of Book I. Looking back, we see Euclid had accomplished much. Even though he had yet to use his parallel postulate, he had nonetheless established four congruence schemes, investigated isosceles triangles, vertical angles, and exterior angles, and had carried out various constructions. But he had gone about as far as he could. The notion of parallels was about to enter the *Elements.*

Book I: Parallelism and Related Topics

PROPOSITION I.27 If a straight line falling on two straight lines make the alternate angles equal to one another, the straight lines will be parallel.

PROOF Here, assuming that $\angle 1 = \angle 2$ in Figure 2.10, Euclid had to establish that lines *AB* and *CD* were parallel—that is, according to Definition 23 he had to prove that these lines can never meet. Adopting an indirect argument, he assumed they intersected and sought a contradiction. That is, suppose *AB* and *CD*, if extended far enough, meet at point *G.* Then the figure *EFG* is a long, stretched-out triangle. But $\angle 2$, an exterior angle of $\triangle EFG$, equals $\angle 1$, an opposite and interior angle of this same triangle. Again, this is impossible according to I.16, the exterior angle theorem. Hence we conclude that *AB* and *CD* never intersect, no matter how far they are extended. Since this is precisely Euclid's definition of these lines' being parallel, the proof is complete.

Q.E.D.

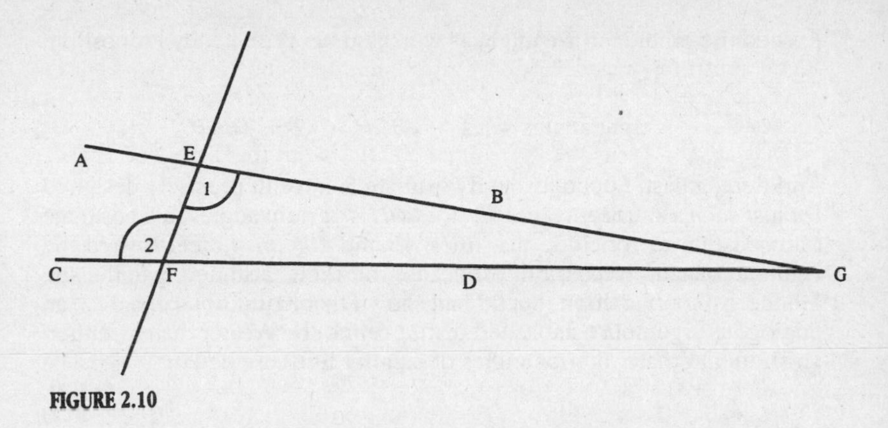

FIGURE 2.10

Proposition I.27 broke the ice with regard to parallelism, but Euclid had yet again avoided the parallel postulate. This last, most controversial postulate finally made its appearance when Euclid proved the converse of I.27 in Proposition I.29.

PROPOSITION I.29 A straight line falling on parallel straight lines makes the alternate angles equal to one another.

PROOF This time, Euclid assumed that *AB* and *CD* in Figure 2.11 are parallel lines and asserted that $\angle 1 = \angle 2$. Again, his mode of attack was indirect. That is, he supposed that $\angle 1 \neq \angle 2$ and from there derived a logical contradiction. For, if these angles were not equal, then one must

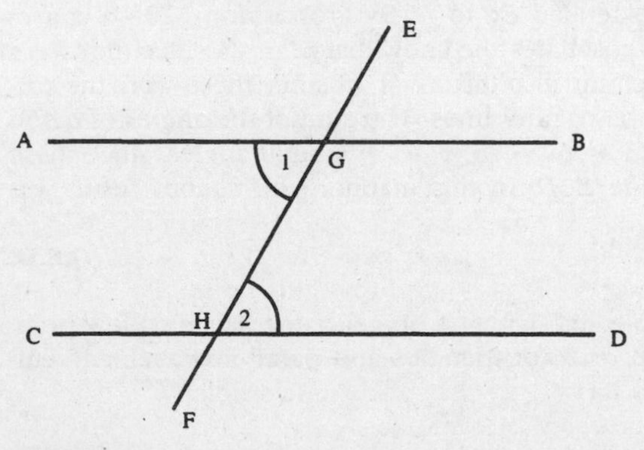

FIGURE 2.11

exceed the other, and we might as well assume $\angle 1 > \angle 2$. By Proposition I.13

$$2 \text{ right angles} = \angle 1 + \angle BGH > \angle 2 + \angle BGH$$

And here, at last, Euclid invoked Postulate 5, a result precisely designed for just such a situation. Since $\angle 2 + \angle BGH < 2$ right angles, his postulate allowed him to conclude that lines *AB* and *CD must* meet toward the right, a blatant impossibility because of their assumed parallelism. Hence, by contradiction, Euclid had shown that $\angle 1$ cannot exceed $\angle 2$; an analogous argument established that $\angle 2$ cannot be greater than $\angle 1$ either. In short, alternate interior angles of parallel lines are equal.

Q.E.D.

As a corollary to the proof, Euclid easily deduced that the corresponding angles were likewise equal, that is, in Figure 2.11, $\angle EGB = \angle 2$. This followed since $\angle EGB$ and $\angle 1$ were vertical angles.

Having at last indulged in the parallel postulate, Euclid now found it virtually impossible to break the habit. Of the remaining 20 propositions in Book I, all but one either used the postulate directly or used a proposition predicated upon it, the lone exception being Proposition I.31, in which Euclid showed how to construct a parallel to a given line through a point not on the line. But the parallel postulate was certainly embedded in the theorem everyone had been waiting for:

PROPOSITION I.32 In any triangle . . . the three interior angles . . . are equal to two right angles.

PROOF Given $\triangle ABC$ in Figure 2.12, he drew *CE* parallel to side *AB* by Proposition I.31 and extended *BC* to *D*. By Proposition I.29—a consequence of the parallel postulate—he knew that $\angle 1 = \angle 4$ since they were alternate interior angles and also that $\angle 2 = \angle 5$ since these were the corresponding angles of the parallel lines. The sum of the angles of $\triangle ABC$ was thus $\angle 1 + \angle 2 + \angle 3 = \angle 4 + \angle 5 + \angle 3 = 2$ right angles, since these formed the straight line *BCD*. In this manner, the famous result was proved.

Q.E.D.

From here, Euclid set his sights on bigger game. His next few propositions dealt with the *areas* of triangles and parallelograms, and culminated in Proposition I.41.

PROPOSITION I.41 If a parallelogram have the same base with a triangle and be in the same parallels, the parallelogram is double of the triangle.

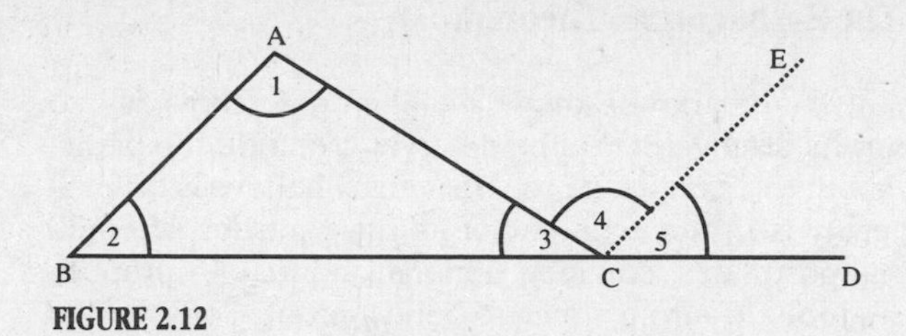

FIGURE 2.12

This was the Greek way of saying that a triangle's area is half that of any parallelogram sharing the triangle's base and having the same height. Since one such parallelogram is a rectangle and since the rectangle's area is (base) × (height), we see that I.41 contained the modern formula Area (triangle) = ½*bh*.

However, Euclid did not think in such algebraic terms. Rather, he envisioned $\triangle ABC$ as literally having the same base as parallelogram *ABDE* and falling between the parallels *AB* and *DE*, as shown in Figure 2.13. Then, as Euclid proved, Area (parallelogram *ABDE*) = 2 Area ($\triangle ABC$).

A few propositions later, in I.46, Euclid showed how, given a line segment, to construct upon it a square. A square, of course, is a regular quadrilateral, since all of its sides and all of its angles are congruent. At first, this may sound like a trivial proposition, especially when one recalls that Book I began with the construction of an equilateral triangle, the regular three-sided figure. Yet a look at his proof shows why this had to be so long delayed. Much of the argument rested on properties of parallels, and these of course had to await the critical I.29. So, whereas Euclid constructed regular *triangles* at the outset of Book I, he waited until nearly the end to do regular *quadrilaterals*.

With these 46 propositions proved, Book I had but two to go. It appears that Euclid had saved the best for last. After all of these preliminaries, he was ready to tackle the Pythagorean theorem, surely one of the most significant results in all of mathematics.

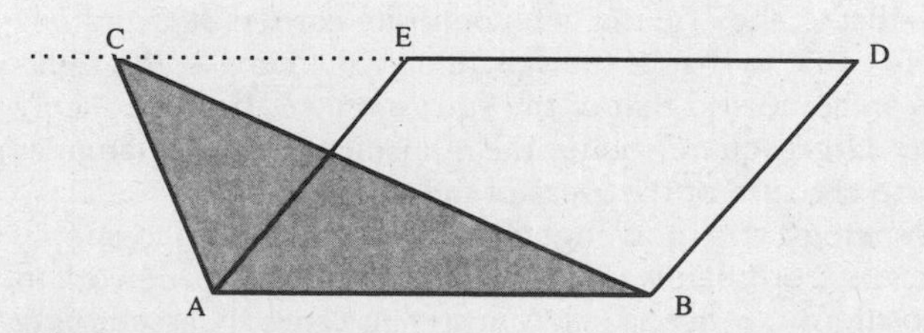

FIGURE 2.13

Great Theorem: The Pythagorean Theorem

As already noted, this landmark was known well before Euclid's day, so he was by no means its discoverer. Yet he deserves credit for the particular proof we are about to examine, a proof that many believe is original with Euclid. Its beauty is in the economy of its prerequisites; after all, Euclid had only his postulates, common notions, and first 46 propositions—a rather lean tool-kit—from which to build a proof. Consider the topics in geometry that he had not yet addressed: the only quadrilaterals he had investigated were parallelograms; circles, by and large, were yet unexplored; and the highly important subject of similarity would not be mentioned until Book VI. It is surely possible to devise short proofs of the Pythagorean theorem by using similar triangles, but Euclid was unwilling to put off the proof of this major proposition until Book VI. He clearly wanted to reach the Pythagorean theorem as early and directly as possible, and thus he devised a proof that would become only the 47th proposition of the *Elements*. In this light, one can see that much of what preceded it pointed toward the great theorem of Pythagoras, which served as a fitting climax to Book I.

Before we plunge into the details, here is the result stated in Euclid's words and a preview of the very clever strategy he used to prove it:

PROPOSITION I.47 In right-angled triangles, the square on the side subtending the right angle is equal to the squares on the sides containing the right angle.

Note that Euclid's proposition was not about an *algebraic* equation $a^2 = b^2 + c^2$, but about a *geometric* phenomenon involving literal squares constructed upon the three sides of a right triangle. Euclid had to prove that the combined areas of the little squares upon AB and AC equaled the area of the large square constructed upon the hypotenuse BC, seen in Figure 2.14. To do this, he hit upon the marvelous strategy of starting at the vertex of the right angle and drawing line AL parallel to the side of the large square and splitting the large square into two rectangular pieces. Then Euclid proved the remarkable fact that the left-hand rectangle—that is, the one that with opposite corners at B and L—had area precisely equal to that of the square on AB; likewise, the right-hand rectangle's area equaled that of the square on AC. It immediately followed that the large square, being the sum of the two rectangular areas, was likewise the sum of the areas of the squares!

This general strategy was most ingenious, but it remained to supply the necessary details. Fortunately, Euclid had done all the spadework in his earlier propositions, so it was just a matter of carefully assembling the pieces.

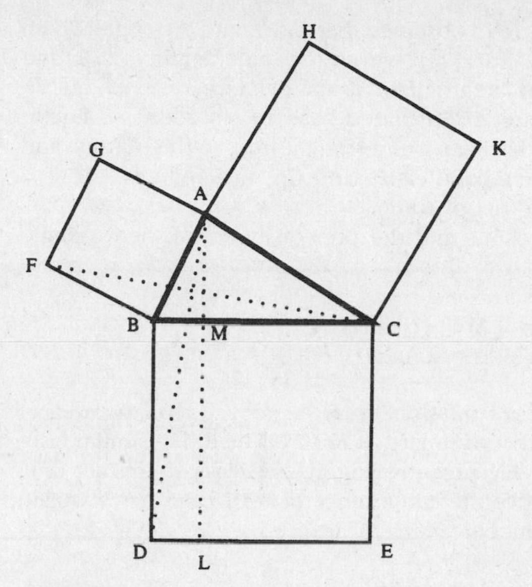

FIGURE 2.14

PROOF By assumption, Euclid knew that $\angle BAC$ was a right angle. He applied I.46 to construct the squares on the three sides, and used I.31 to draw *AL* through *A* and parallel to *BD*. He then drew lines *AD* and *FC*, for reasons that may at first appear mystifying but which should soon become apparent.

It was critical for Euclid to establish that *CA* and *AG* lie on the same straight line. This he did by noting that $\angle GAB$ was right by the construction of the square, while $\angle BAC$ was right by hypothesis. Since these two angles sum to two right angles, Proposition I.14 guaranteed that *GAC* was itself a straight line. Interestingly, it was in proving this apparently minor technical point that he made his one and only use of the fact that $\angle BAC$ is right.

Now Euclid looked at the two slender triangles *ABD* and *FBC*. Their shorter sides—*AB* and *FB*, respectively—were equal since they were the sides of a square; their longer sides—*BD* and *BC*—were equal for the same reason. And what about the corresponding included angles? Notice that $\angle ABD$ is the sum of $\angle ABC$ and the square's right angle $\angle CBD$, while $\angle FBC$ is the sum of $\angle ABC$ and the square's right angle $\angle FBA$. Postulate 4 stipulates that all right angles are equal, Common Notion 2 guarantees that sums of equals are equal, and thus $\angle ABD = \angle FBC$. Consequently, Euclid had established that the narrow triangles *ABD* and *FBC* were congruent by SAS—that is, by I.4; hence they have the same areas.

So far, so good. Next Euclid observed that $\triangle ABD$ and rectangle *BDLM* shared the same base (*BD*) and fell within the same parallels (*BD* and *AL*), and thus by I.41, the area of *BDLM* was twice the area of $\triangle ABD$. Similarly, $\triangle FBC$ and square *ABFG* shared base *BF*. In addition, Euclid had taken pains to prove that *GAC* was a straight line, so the triangle and the square both lay between parallels *BF* and *GC*; again by I.41, the area of square *ABFG* was twice that of $\triangle FBC$.

He combined these results and the previously established triangle congruence to conclude:

$$\begin{aligned}\text{Area (rectangle } BDLM) &= 2 \text{ Area } (\triangle ABD)\\ &= 2 \text{ Area } (\triangle FBC) = \text{Area (square } ABFG)\end{aligned}$$

This was half of Euclid's mission. Next he proved that the area of rectangle *CELM* equaled that of square *ACKH*. This he did in similar fashion, first drawing *AE* and *BK*, next proving that *BAH* was a straight line, and then using SAS to prove the congruence of $\triangle ACE$ and $\triangle BCK$. Again applying I.41, Euclid deduced:

$$\begin{aligned}\text{Area (rectangle } CELM) &= 2 \text{ Area } (\triangle ACE)\\ &= 2 \text{ Area } (\triangle BCK) = \text{Area (square } ACKH)\end{aligned}$$

Finally, the Pythagorean theorem was at his fingertips, for

$$\begin{aligned}&\text{Area (square } BCED)\\ &\quad = \text{Area (rectangle } BDLM) + \text{Area (rectangle } CELM)\\ &\quad = \text{Area (square } ABFG) + \text{Area (square } ACKH)\end{aligned}$$

Q.E.D.

Thus ended one of the most significant proofs in all of mathematics. The diagram Euclid used (Figure 2.14) has become justly famous. It is often called "the windmill" because of its resemblance to such a structure. The windmill is evident on the page shown here, written in Latin, from a 1566 edition of the *Elements*. Obviously, students over four centuries ago were grappling with this figure even as we ourselves have just done.

Euclid's proof is not, of course, the only way to establish the Pythagorean theorem. There are, in fact, hundreds of different versions, ranging from the highly ingenious to the dreadfully uninspired. (These include a proof devised by an Ohio Congressman named James A. Garfield, who went on to the U.S. presidency.) Readers interested in sampling other arguments may wish to consult E. S. Loomis' book *The Pythagorean*

PROPOSIT. XLVII.

Theorema.

ΕΝ τοῖς ὀρθογωνίοις τριγώνοις· τὸ ἀπὸ τῆς τὴν ὀρθὴν γωνίαν ὑποτεινούσης πλευρᾶς τετράγωνον ἴσον ἐστὶ, τοῖς ἀπὸ τῶν τὴν ὀρθὴν γωνίαν περιεχουσῶν πλευρῶν τετραγώνοις.

In triangulis rectangulis: quadratum lateris angulum rectum subtendentis, est æquale quadratis laterum, rectum angulum continentium.

ἡ ἔκθεσις.

Sit triangulus rectangulus αβγ, habens angulum βαγ rectum. ὁ διορισμός. Dico quod quadratum lateris βγ, est æquale quadratis laterum βα, αγ. ἡ κατασκευή. Describatur à linea βγ, quadratum βδεγ. Item à linea βα quadratum βη. Praeterea à linea αγ quadratum γθ. Ducatur per punctum α, alterutri linearum βδ, γε, æquedistans recta linea αλ. Ducantur duæ lineæ rectæ αδ, ζγ.

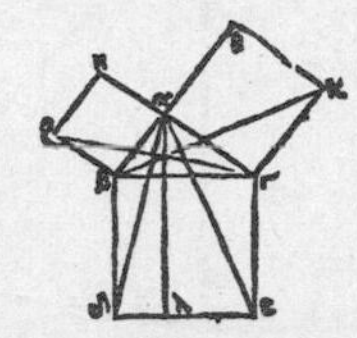

Proposition I.47 from 1566 edition of the *Elements* (photograph courtesy of The Ohio State University Libraries)

Proposition for a bewildering, if not mind-numbing, collection of hundreds upon hundreds of proofs of this remarkable theorem.

Proposition I.47 marked the high point of Book I, but Euclid had a final result to prove, the converse of the Pythagorean theorem. Here again Euclid's ingenuity and economy were undeniable. Unfortunately, this proof is not so well known as it should be. In fact, while most students encounter a proof of the Pythagorean theorem at some point in their lives, far fewer see a proof of the converse or, for that matter, are even sure of its validity.

Two features of the proof deserve special mention. First, it is quite short, especially when compared to the argument we have just seen. Second, Euclid *used* the Pythagorean theorem in establishing its converse. While not unprecedented, this logical approach is at least worthy of note. Recall that in proving the two great propositions about parallels—I.27 and its converse I.29—Euclid did not use one in the proof of the other. His approach to the converse of the Pythagorean theorem, however, made them a definite sequential unit, with I.48 resting firmly upon the foundation of I.47.

PROPOSITION I.48 If in a triangle the square on one of the sides be equal to the squares on the remaining two sides of the triangle, the angle contained by the remaining two sides of the triangle is right.

PROOF Euclid began with $\triangle ABC$ and assumed that $\overline{BC}^2 = \overline{AB}^2 + \overline{AC}^2$, as shown in Figure 2.15. He had to prove that $\angle BAC$ was a right angle.

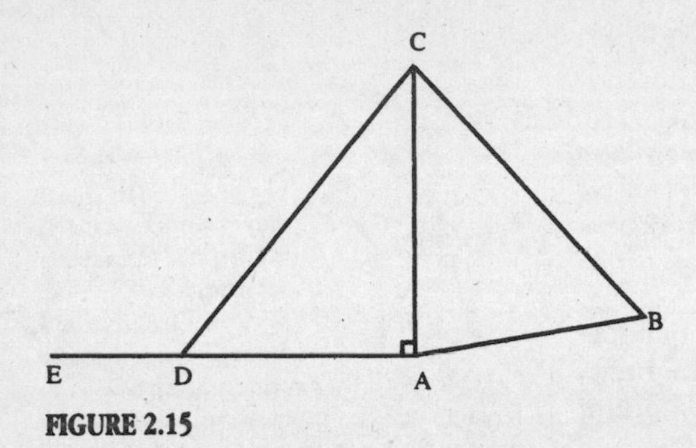

FIGURE 2.15

To do this, Euclid first drew AE perpendicular to AC at A, by Proposition I.11. He then constructed $\overline{AD} = \overline{AB}$, and drew CD. The heart of his argument now lay in proving that triangles BAC and DAC were congruent.

Clearly, the two triangles shared side AC, and $\overline{AD} = \overline{AB}$ by construction. While we obviously cannot assert that $\angle BAC$ is right (in fact, that is what the theorem is trying to establish), we do know that $\angle DAC$ is right by the construction of the perpendicular. Thus, Euclid was perfectly justified in applying the Pythagorean theorem to right triangle DAC to deduce that

$$\overline{CD}^2 = \overline{AD}^2 + \overline{AC}^2 = \overline{AB}^2 + \overline{AC}^2 = \overline{BC}^2 \qquad \text{by hypothesis}$$

But the equality of $\overline{CD}^2$ and $\overline{BC}^2$ implies the equality of $\overline{CD}$ and $\overline{BC}$ as well. Hence ΔDAC and ΔBAC are congruent by SSS. As a consequence, $\angle BAC$ and $\angle DAC$ must be congruent. But the latter was constructed to be a right angle. Thus $\angle BAC$ is a right angle as well.

Q.E.D.

Taken in tandem, Propositions I.47 and I.48 completely characterize right triangles. Euclid has shown that a triangle is right *if and only if* the square on the hypotenuse equals the sum of the squares on the legs. These proofs were, and remain, an example of geometry at its best.

Yet these two Pythagorean propositions are remarkable in another sense. It is one thing for Euclid to have proved them in such a fine fashion, but it is another thing for them to be true in the first place. There is no intuitive reason that right triangles should have such an intimate connection to the sums of squares. Unlike I.20, for instance, this is not a theorem whose truth is evident even to an ass. On the contrary, the Pythagorean theorem establishes a supremely odd fact, one whose oddness is unrecognized only because the result is so famous. This intrinsic strangeness was well expressed by Richard Trudeau in his book *The Non-Euclidean Revolution.* Trudeau observed that right angles are familiar, everyday entities that appear not only in the man-made world but also in Nature itself. What could be more "ordinary" or "natural" than right angles? And yet, says Trudeau:

> To me the Theorem of Pythagoras is very surprising . . . . "$a^2 = b^2 + c^2$" . . . evokes no visceral memories whatever . . . Because the equation is abstract and precise, it is alien. I can't imagine what such a thing could possibly have to do with everyday right angles. So, when the pall of familiarity lifts, as it occasionally does, and I see the Theorem of Pythagoras afresh, I am flabbergasted.

Epilogue

Down through history, the most troubling feature of Book I of the *Elements* was the controversial parallel postulate. The trouble arose not because anyone doubted that the parallel postulate had to be true. On the contrary, it was universally agreed that the postulate was a logical necessity. After all, geometry was an abstract way of describing the universe—a kind of "pure physics"—and surely physical reality dictated the truth of the parallel postulate.

Thus, it was not the necessity of Euclid's statement that was challenged. Rather it was its classification as a *postulate* rather than as a proposition. The classical writer Proclus summed up this view with his comment, "This [Postulate 5] ought even to be struck out of the Postulates altogether; for it is a theorem. . . ."

This conviction was not surprising. First of all—and this may have really bothered the ancient geometers—the postulate *sounded* like a proposition, for its statement consumed the better part of a paragraph. Moreover, not only did Euclid seem to avoid using the postulate as long as he could, but he managed to prove some fairly sophisticated results without it. "If his other postulates and common notions were rich enough to yield such theorems as I.16 or I.27 or four different congruence schemes," so the reasoning went, "then surely they should likewise imply the parallel postulate."

For what appeared to be very good reasons, the search was on for a derivation of Postulate 5. In seeking such a proof, mathematicians were free to use any of the other postulates or common notions, as well as Euclid's propositions I.1 through I.28, none of which involved the statement in question. Uncounted mathematicians tried their hand at concocting a proof. Unfortunately, years of frustration became decades and then centuries of failure. The proof remained elusive.

What geometers did in the process was to find a host of new results logically equivalent to the parallel postulate. It often happened that a purported proof of Postulate 5 required the mathematician to assume a seemingly obvious but hitherto unproven statement. Unfortunately—and here lay the problem—the parallel postulate itself was necessary in order to derive this statement. To a logician, this says that both were really expressing the identical concept, and a "proof" of Postulate 5 that required the assumption of its logical equivalent was of course no proof at all.

Four of the more famous equivalents of the parallel postulate appear below. It should be stressed that, had any one of these been proved from Postulates 1 through 4, then Postulate 5 would likewise follow.

- **Proclus' axiom**: If a line intersects one of two parallels, it must intersect the other also.
- **equidistance postulate**: Parallel lines are everywhere equidistant.
- **Playfair's postulate**: Through a point not on a given line, there can be drawn one and only one line parallel to the given line.
- **the triangle postulate**: The sum of the angles of a triangle is two right angles.

These logical equivalents notwithstanding, the nature of the parallel postulate remained unresolved through the Renaissance. Whoever deduced the parallel postulate would have been guaranteed everlasting fame in the annals of mathematics. At times the proof seemed tantalizingly close, yet it evaded the efforts of the world's finest mathematical minds.

Then, early in the nineteenth century, three mathematicians simultaneously had the burst of insight necessary to see the matter in its true light. The first was the incomparable Carl Friedrich Gauss (1777–1855), whose biography is delayed until Chapter 10. Gauss recast the issue in terms of the degree-measure of the angles of a triangle. Wanting to prove that triangles must contain 180°, he assumed for the sake of argument that they did not. This left him with two alternatives: that triangles have more than 180° in their angles or that they have less. He proceeded to investigate these two cases.

Using the fact that lines are infinitely long (an assumption that Euclid likewise had made implicitly and that no one to that point had challenged), Gauss found that a triangle's angle sum exceeding 180° led to a logical contradiction. Thus, that case was effectively eliminated. If he could likewise dispense with the other case, he would have established, indirectly, the necessity of the parallel postulate.

Beginning with the assumption that triangles have fewer than 180° in their angles, Gauss started deriving consequences. These turned out to be quite strange, seemingly bizarre and counter-intuitive (one is presented shortly). Yet nowhere did Gauss find the *logical* contradiction he sought. In 1824, he summarized the situation by stating:

> . . . that the angle sum of a triangle can't be less than 180° . . . this is . . . the reef on which all the wrecks occur.

Gradually, as Gauss delved more and more deeply into this peculiar geometry, he became convinced that no logical contradiction existed. Rather, he began to sense that he was developing not an *inconsistent* geometry but just an *alternative* one, a "non-Euclidean" geometry, in his words. Gauss said as much in a private letter of 1824:

> The assumption that the sum of the three angles is less than 180° leads to a curious geometry, quite different from ours, but thoroughly consistent, which I have developed to my entire satisfaction.

This was a breathtaking statement. Yet Gauss, universally regarded as the foremost mathematician of his day, did not publicize his findings. Perhaps the burdens of fame figured in his decision, for he was certain the controversial nature of his position would cause an uproar that might jeopardize his lofty reputation. In an 1829 letter to a confidant, Gauss observed that he had no plans

> . . . to work up my very extensive researches for publication, and perhaps they will never appear in my lifetime, for I fear the howl of the Boeotians if I speak my opinion out loud.

While today's reader may miss a bit of this classical allusion, suffice it to say that being called a "Boeotian" is being labeled an unimaginative, crudely obtuse dullard. Obviously Gauss had little regard for the receptivity of the mathematical community to his new ideas.

Next entered the Hungarian mathematician Johann Bolyai (1802–1860). Johann's father Wolfgang had been an associate of Gauss and had himself spent much of a lifetime in a futile attempt to prove Euclid's postulate. In an age when sons often took the professions of their fathers—be they clergymen or cobblers or chefs—we have here the younger Bolyai taking from his father the rather esoteric career of trying to derive the parallel postulate. Wolfgang, however, knew all too well the difficulties of such a career and wrote this strong warning to his son:

> You must not attempt this approach to parallels. I know this way to its very end. I have traversed this bottomless night, which extinguished all light and joy of my life . . . . I entreat you, leave the science of parallels alone.

The young Johann Bolyai did not heed his father's advice. Much like Gauss, Johann came to recognize the crucial trichotomy involving the angle-sum of a triangle and tried to eliminate all but the case equivalent to the parallel postulate; like Gauss, he was unsuccessful. As Bolyai delved ever more deeply into the problem, he too arrived at the conclusion that Euclid's geometry had a logically valid competitor, and wrote in astonishment at his peculiar yet apparently consistent propositions, "Out of nothing, I have created a strange new universe."

Unlike Gauss, Johann Bolyai was not reluctant to publish his findings, and these appeared as an appendix to an 1832 work by his father. The elder Bolyai enthusiastically sent a copy of the book to his friend Gauss; father and son could only have been surprised by Gauss' response:

> If I begin with the statement that I dare not praise [your son's] work, you will of course be startled for a moment: but I cannot do otherwise; to praise it would amount to praising myself; for the entire content of the work, the path which your son has taken, the results to which he is led, coincide almost exactly with my own meditations which have occupied my mind for from thirty to thirty-five years.

It is easy to see that Gauss hit his enthusiastic young admirer with a blast of cold water. To his credit, Gauss graciously described himself to be ". . . overjoyed that it happens to be the son of my old friend who outstrips me in such a remarkable way." Still, for Johann to learn that his greatest discovery had been sitting in Gauss' drawer for decades came as a severe blow to his ego.

But Johann's ego had one more trial to endure, for it soon came to light that a Russian mathematician, Nikolai Lobachevski (1793–1856), not only had traveled the same path as Gauss and Bolyai, but had published his own account of non-Euclidean geometry in 1829—a full three years before. Lobachevski, however, had written his treatise in Russian, and it apparently had gone unnoticed in western Europe. We have here a phenomenon not uncommon in science, that of a discovery made simultaneously and independently by many individuals. As Wolfgang Bolyai so charmingly observed:

> . . . it seems to be true that many things have, as it were, an epoch in which they are discovered in several places simultaneously, just as the violets appear on all sides in springtime.

The impact of these discoveries had barely struck home when yet another innovator, Georg Friedrich Bernhard Riemann (1826–1866), adopted a different viewpoint about the infinite length of geometric lines. It had been this infinitude that had allowed Gauss, Bolyai, and Lobachevski to eliminate the case in which triangles contained more than 180°. But was there a need to assume this infinitude at all? Euclid's second postulate asserted that a straight line could be continued in a straight line, but was this not asserting simply that one never reached the end of a line? Riemann could easily imagine the case where lines—somewhat like circles—are of finite length yet have no "end." He put it this way:

> . . . we must distinguish between *unboundedness* and *infinite extent*. . . . The unboundedness of space possesses . . . a greater empirical certainty than any external experience. But its infinite extent by no means follows from this.

When Riemann reexamined geometry under the assumption of unbounded but finite lines, the contradiction to a triangle's exceeding 180° disappeared. Consequently, he developed another kind of non-Euclidean geometry, one in which the angles of a triangle sum to more than two right angles. Although different from both Euclid's and Bolyai's, Riemann's geometry was apparently just as consistent.

Today, we recognize all four of these individuals as the originators of non-Euclidean geometry. It seems fair that, as pioneers, they should share the glory. But even their discoveries did not fully resolve the fundamental issue of the parallel postulate. For, while they had developed their geometries to a high level of sophistication, it was merely a feeling in their bones, not a logical argument on paper, that supported their contention that the new geometries were valid alternatives to Euclid's. In spite of the strong convictions of Gauss, Bolyai, Lobachevski, and Riemann, the possibility remained that at some point in the future, a brilliant mathematician might yet derive a contradiction from the assumption that the angle sum of a triangle was less than, or more than, 180°.

Thus, the final chapter of this age-old story was written in 1868 by the Italian Eugenio Beltrami (1835–1900), who unequivocally proved that non-Euclidean geometry was as logically consistent as Euclid's own. That is, if a contradiction lurked somewhere in the geometry of Gauss, Bolyai, and Lobachevski, or in that of Riemann, then Beltrami showed that a contradiction also had to exist in the geometry of Euclid. Since virtually everyone felt that Euclid's geometry was as consistent as could be, the conclusion was that non-Euclidean geometries were likewise as good as gold. Put another way, non-Euclidean geometry is not *logically* inferior to its older, Euclidean counterpart.

To get some idea of the strange content of the Gauss/Bolyai/Lobachevski brand of non-Euclidean geometry—that is, the kind where triangles have fewer than 180° in their angles—consider the proofs of a pair of non-Euclidean results. The first involves another look at the congruence of triangles. Of course, Euclid's congruence schemes, established prior to his first use of Postulate 5, remain valid in this non-Euclidean realm since they were proved without reference to anything but his other postulates and common notions. The surprising development is that, in Bolyai's geometry, there is yet another way to show congruence, namely "angle-angle-angle."

In Euclid's geometry, when two triangles have their angles respectively equal, we know the triangles are similar. They would have the same *shape* but need not be congruent; we could have, for example, a tiny equilateral triangle and a large equilateral triangle, non-congruent figures all of whose angles are equal. The non-Euclidean theorem that follows shows that no such thing is possible in this strange world. If two of Bolyai's triangles have the same shape, they must have the same size!

THEOREM (AAA) If two triangles have the three angles of one respectively equal to the three angles of the other, then the triangles are congruent.

PROOF For triangles *ABC* and *DEF* in Figure 2.16, assume that $\angle 1 = \angle 4$, $\angle 2 = \angle 5$, and $\angle 3 = \angle 6$. We assert that sides *AB* and *DE* *must* have the same length. To prove this, suppose for the sake of an eventual contradiction that they differ in length, and without loss of generality we might as well assume that $\overline{AB} < \overline{DE}$.

Construct $\overline{DG} = \overline{AB}$ and, by I.23, draw $\angle DGH = \angle 2$. It is clear that $\triangle ABC$ and $\triangle DGH$ are congruent by ASA, and it follows that $\angle DGH = \angle 2 = \angle 5$ and similarly $\angle DHG = \angle 3 = \angle 6$.

Now examine the lower quadrilateral *EFHG*. Since *DGE* and *DHF* are straight lines, we know by I.13 that $\angle EGH = (180° - \angle DGH) = (180° - \angle 5)$ and that $\angle FHG = (180° - \angle DHG) = (180° - \angle 6)$. Thus the measures of the four angles of quadrilateral *EFHG* sum to

$$(180° - \angle 5) + (180° - \angle 6) + \angle 6 + \angle 5 = 360°$$

But now draw the diagonal *GF* through this quadrilateral. This divides it into two triangles, each of which has fewer than 180° in its

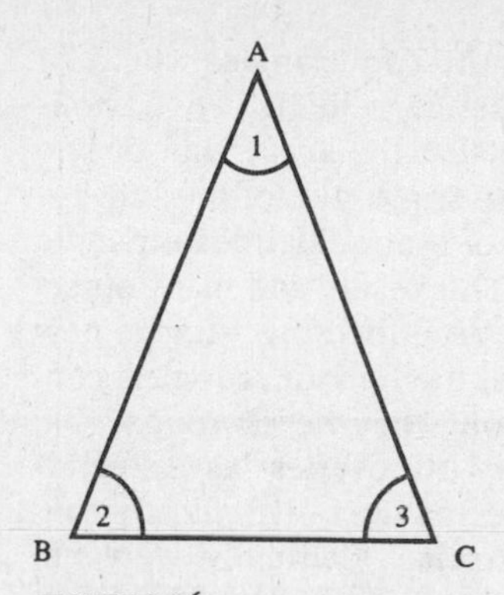

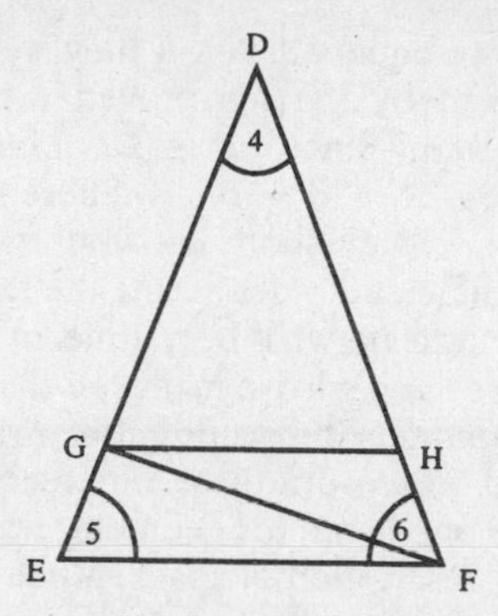

FIGURE 2.16

angles by our basic non-Euclidean property; so the sum of all the angles in the two triangles must be *less than* 360°. Yet this sum is precisely the combined total of the four angles in our quadrilateral, already shown to be *equal to* 360°.

We have reached a contradiction. This means that the first step, in which we assumed $\overline{AB} \neq \overline{DE}$, was erroneous. In short, these sides are equal in length. But then we get immediately that the original triangles *ABC* and *DEF* are congruent by ASA—that is, by Proposition I.26—which is what we set out to show.

Q.E.D.

It is easy to draw a surprising corollary from this proposition: In non-Euclidean geometry, not all triangles have the same angle sum! This most fundamental property of Euclid's geometry—one that figured prominently in so much geometric reasoning—must be discarded when we move to the non-Euclidean domain. For suppose we consider the two triangles shown in Figure 2.17, each having angles α and β at their base, but with side *AB* much shorter than side *DE*. Now we assert that

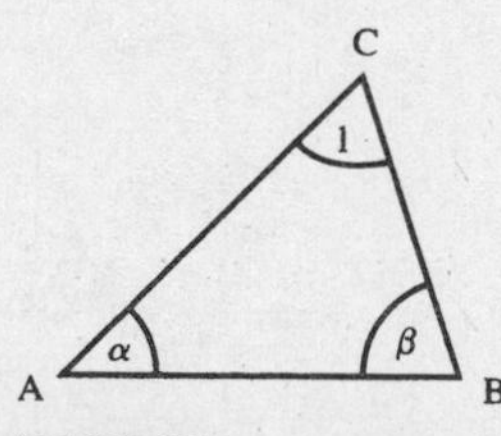

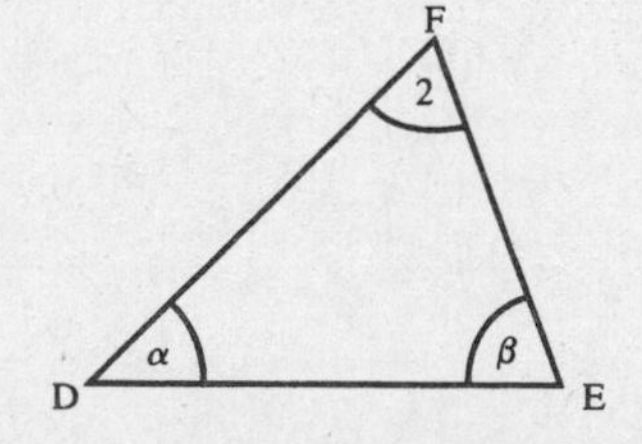

FIGURE 2.17

$\angle 1$ cannot equal $\angle 2$. For, if they were equal, the two triangles would be congruent by the just-proved AAA congruence scheme, an obvious impossibility since $\overline{AB} \neq \overline{DE}$. Thus we see that the angle sum of one triangle—$\angle 1 + \alpha + \beta$—is different from the angle sum of the other—$\angle 2 + \alpha + \beta$. In short, knowing two angles of a non-Euclidean triangle is not sufficient to determine the third one. This result, and many others like it, indicate what Bolyai meant when he described his "strange new universe" and why so many felt that, just over the horizon, a logical contradiction must be waiting. But as it turned out, they were wrong.

And where did these nineteenth century discoveries leave Euclid? On the one hand, his geometry was displaced as the only logically consistent description of space. Much to the surprise of virtually everyone, it turned out that the parallel postulate was *not* mandated by logic. Euclid assumed it, but there was no mathematical necessity to do so. Competing geometries, equally as valid, existed.

Yet the net effect may be to enhance, not destroy, Euclid's reputation. For he, unlike so many who followed, did not fall into the trap of trying to prove the parallel postulate from the other self-evident truths, an endeavor, we now know, that is utterly doomed to failure. Instead, he simply laid out his assumption where it properly belonged, as a postulate. Euclid certainly could not have known about the alternative geometries that would be discovered two millennia in the future. Yet something in his mathematician's intuition must have told him that this property was a separate, independent idea that needed its own postulate, no matter how wordy and complicated it sounded. Mathematicians 22 centuries later proved that Euclid had been right all along.

3
Chapter

Euclid and the Infinitude of Primes
(ca. 300 B.C.)

The *Elements*, Books II–VI

The 48 propositions of Book I of the *Elements* stand as a monument to Euclid's mathematical and organizational skills. Because it comes first, this book is certainly the best known and most studied part of the *Elements*, but it is just one of 13 books into which the work was divided. Chapter 3 offers a quick tour of the rest of this classic text.

Book II explored what we now call "geometric algebra." That is, it framed in geometric terms certain relationships that now are most easily translated into algebraic equations. Of course, the notion of algebra was foreign to the Greeks, and its appearance as a formal system lay centuries in the future. We can get a sense of Book II by citing a representative proposition, one whose statement at first glance appears rather convoluted, but which upon closer examination emerges as a simple and well-known algebraic formula.

PROPOSITION II.4 If a straight line be cut at random, the square on the whole is equal to the squares on the segments and twice the rectangle contained by the segments.

PROOF Euclid began with line segment *AB*, cut at an arbitrary point *C*, as shown in Figure 3.1. If we let $\overline{AC} = a$ and $\overline{BC} = b$, then it is geometrically obvious that the area of "the square on the whole"—that is, $(a + b)^2$—equals the sum of the areas of the squares on the two segments—$a^2 + b^2$—plus twice the area of the rectangle formed by the two segments—$2ab$. In other words,

$$(a + b)^2 = a^2 + b^2 + 2ab = a^2 + 2ab + b^2$$

Q.E.D

This, of course, is a famous identity encountered in the first year of algebra. Euclid approached it not as some algebraic expression but as a literal geometric decomposition of the square upon *AB* into two smaller squares and two congruent rectangles. Yet the equivalence of his geometric statement and its algebraic counterpart is clear. Much of Book II was of this nature. It concluded with Proposition II.14, addressing the quadrature of general polygons, whose proof was examined in Chapter 1.

The third book contained 37 propositions about circles. Circles had been used in the constructions of Book I but had not themselves been the focus of the discussion. In Book III, Euclid proved the standard results about chords, tangents, and angles in circles. Proposition III.1 showed how to find the center of a given circle. Of course, by Definition 15, every circle *has* a center, but for a circle already drawn upon the page, it is not immediately clear how to find that central point. Thus, Euclid provided the necessary construction.

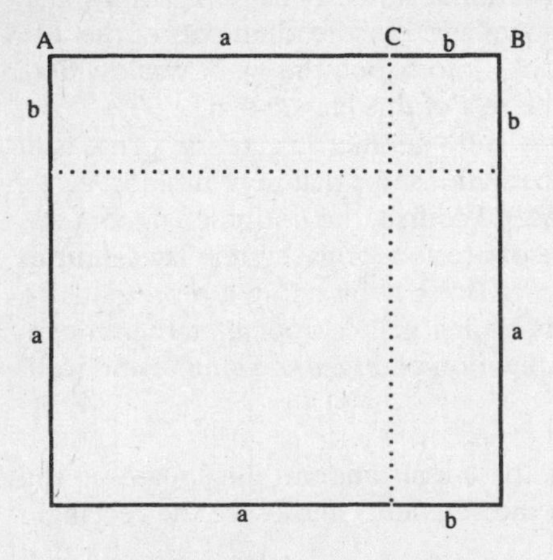

FIGURE 3.1

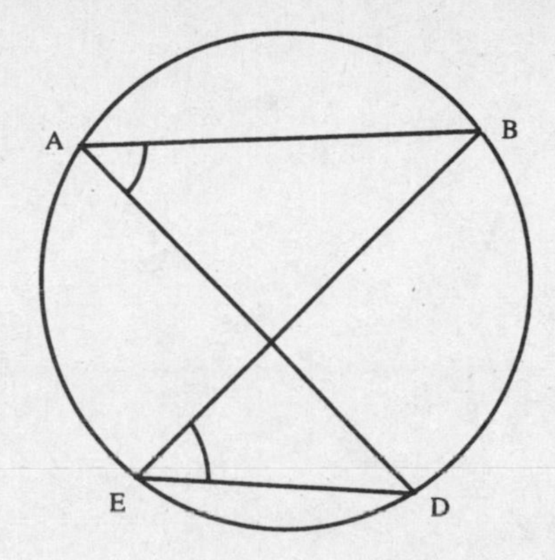

FIGURE 3.2

In Proposition III.18, Euclid gave a clever argument to prove that a tangent to a circle and the radius drawn to that point of tangency meet at right angles. A few propositions later, we find the important result, "In a circle, angles in the same segment are equal to one another." That is, in Figure 3.2 $\angle BAD$ and $\angle BED$ are congruent since both are contained in the segment of the circle *BAED*. In modern terminology, we would say that both intercept the same arc, namely, arc *BD*.

Having proved this theorem, Euclid tackled the concept of a quadrilateral inscribed within a circle, a figure often called a "cyclic quadrilateral." Although this result may appear somewhat specialized, it will figure prominently in the great theorem of Chapter 5, and thus Euclid's simple proof is included here.

PROPOSITION III.22 The opposite angles of quadrilaterals in circles are equal to two right angles.

PROOF We begin with cyclic quadrilateral *ABCD* and draw the two diagonals *AC* and *BD*, as shown in Figure 3.3. Note that $\angle 1 + \angle 2 + \angle DAB =$ 2 right angles, since these are the angles of $\triangle ABD$. But $\angle 1 = \angle 3$ since both intercept arc *AD*; and $\angle 2 = \angle 4$ since both intercept arc *AB*. Hence

$$\begin{aligned} \text{2 right angles} &= (\angle 1 + \angle 2) + \angle DAB \\ &= (\angle 3 + \angle 4) + \angle DAB = \angle DCB + \angle DAB \end{aligned}$$

In other words, the opposite angles of the cyclic quadrilateral indeed add to two right angles, and the proof is complete.

Q.E.D.

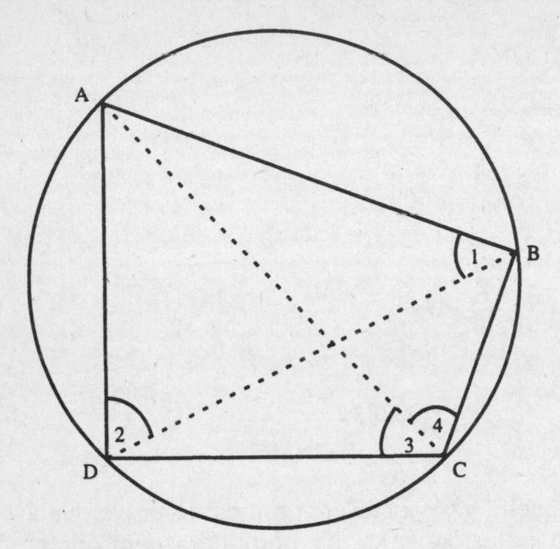

FIGURE 3.3

Later, Proposition III.31 established that an angle inscribed in a semicircle is right, a proof presented in Chapter 1. In that regard, note that nowhere in his book on circles did Euclid address the issue of lunes, nor did Book III contain the familiar results for a circle's circumference ($C = \pi D$) or area ($A = \pi r^2$). A full treatment of these latter topics would have to await the arrival of Archimedes, as discussed in Chapter 4.

Euclid's fourth book dealt with inscribing and circumscribing certain kinds of geometric figures. As with all constructions in the *Elements*, he was limited to his compass and unmarked straightedge. These limitations aside, he nonetheless produced some fairly sophisticated results.

For instance, Proposition IV.4 showed how to *inscribe* a circle within a given triangle, the key being to take as the circle's center the point where the bisectors of the angles of the triangle meet. In the next proposition, he showed how to *circumscribe* a circle about a given triangle; this time, he located the center of the circle at the point where the perpendicular bisectors of the sides meet.

From there, Euclid considered the construction of regular polygons, all of whose sides are the same length and all of whose angles are congruent. These are "perfect" polygons whose symmetry and beauty certainly appealed to the Greek imagination.

Recall that Euclid had begun the *Elements* with the construction of a regular, or "equilateral," triangle, and in Proposition I.46 he had constructed a square on a given segment. In Proposition IV.11, Euclid expanded his repertoire by inscribing a regular pentagon in a circle, and in Proposition IV.15, he inscribed a regular hexagon. The final construction in this book was of the regular pentadecagon—that is, the regular 15-sided polygon—and his argument warrants a quick look.

Within a given circle, Euclid inscribed both an equilateral triangle with side *AC* and a regular pentagon with side *AB*, each sharing a vertex at *A* (Figure 3.4). As Euclid observed, arc *AC* is a third of the circle's circumference, while arc *AB* is a fifth of the same. Consequently, their difference, arc *BC*, intercepts $\frac{1}{3} - \frac{1}{5} = \frac{5}{15} - \frac{3}{15} = \frac{2}{15}$ of the circumference. If we bisect the chord from *B* to *C* and draw the perpendicular outward from the chord's midpoint to point *E* on the circle, we shall have bisected arc *BC*. Thus, arc *BE* is *one*-fifteenth of the circle, and so chord *BE* is the length of the side of a regular pentadecagon. Copying 15 of these chords around the circle completes the construction.

Euclid said no more about regular polygons in the *Elements*, but he clearly was aware that if one had constructed such a polygon, the bisection procedure outlined above would produce regular polygons with twice as many sides. After constructing an equilateral triangle, Greek geometers could therefore produce regular polygons of 6, 12, 24, 48, . . . sides; starting from the square, they could generate regular polygons of 8, 16, 32, 64, . . . sides; from the regular pentagon would emerge regular 10-, 20-, 40-, . . .-gons; and from Euclid's final construction, the pentadecagon, would flow regular polygons of 30, 60, 120, . . . sides.

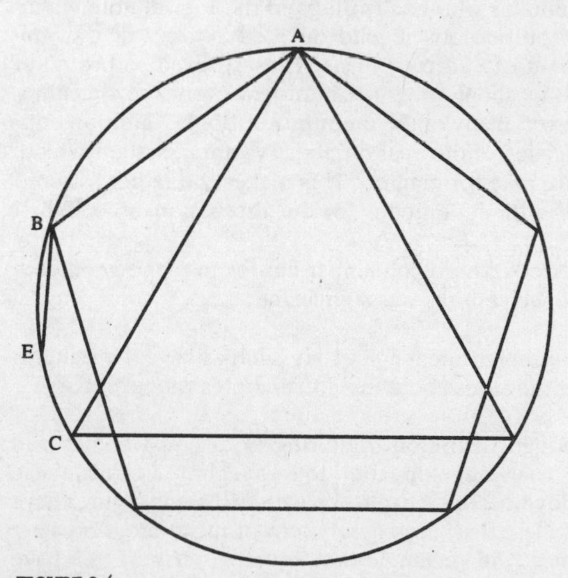

FIGURE 3.4

This was quite a rich collection of constructible regular polygons, but obviously not all regular polygons appeared on this list. Nowhere, for instance, did Euclid mention constructing a regular 7-gon, or 9-gon, or 17-gon, since these did not fit the neat "doubling" patterns above. One imagines that the Greeks put a lot of time and effort into trying to construct other regular polygons, but apparently their efforts led nowhere. In fact, while Euclid did not explicitly say so, most subsequent mathematicians assumed that his were the *only* constructible regular polygons and that any others were simply beyond the capability of compass and straightedge.

It was thus a shock of monumental proportions when the teenaged Carl Friedrich Gauss discovered how to construct a regular heptadecagon (17-gon) in 1796. This discovery marked the young Gauss as a mathematical genius of the first order. Gauss was introduced in the previous chapter with regard to his work in non-Euclidean geometry, and Chapter 10 will return to this remarkably gifted mathematician.

In summary, Books I through IV of the *Elements* addressed the essentials of triangles and polygons, of circles, and of the regular polygons. At this point, Euclid had done about as much geometry as he could without the highly useful notion of similarity. As mentioned in Chapter 1, similarity arguments and the proportions they generate had received a fatal blow with the Pythagorean discovery of incommensurable magnitudes, and it was Eudoxus who finally plugged the logical hole with a satisfactory theory of proportions. Euclid devoted Book V of the *Elements* to a development of Eudoxus' ideas. These proved so profound as to influence thinking about irrational numbers even into the nineteenth century. However, many of the theorems of Book V are now subsumed into the properties of our real number system, a system which, for better or worse, we take for granted. This makes the rather tortured arguments of Book V a bit superfluous for our discussion, so we shall move on to Book VI.

Here Euclid undertook a study of similar figures in plane geometry. His very definition of such figures was significant.

□ **Definition VI.1** *Similar* rectilineal figures are such as have their angles severally equal and the sides about the equal angles proportional.

This was a double-edged definition, requiring both equal angles and proportional sides to guarantee similarity. In less technical terms, these two conditions embody what we mean when we say that two figures have the same shape. It is clear that, in general, both of these properties are necessary. For instance, the rectangle and square in Figure 3.5 have equal angles, but the non-proportionality of their sides distorts them into

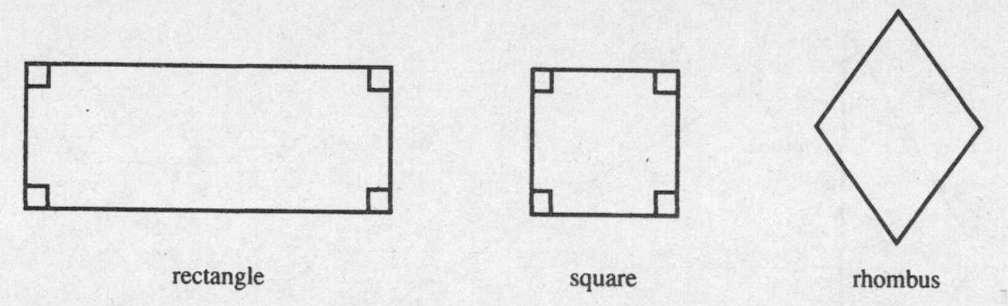

FIGURE 3.5

different shapes. On the other hand, the square and rhombus have their sides in the same proportions—that is, 1:1—but their differing angles give them quite different shapes as well.

Interestingly, these dual requirements for similarity vanish when attention is restricted to the realm of *triangles.* Making use of the Eudoxean theory of Book V, Euclid proved, in Proposition VI.4, that if two triangles have their corresponding angles equal, then their corresponding sides must be proportional; conversely, in Proposition VI.5, he showed that if two triangles have their sides proportional, then their corresponding angles must be equal. In short, the whole matter simplifies greatly for three-sided figures, since either of the two similarity conditions implies the other. Consequently, it comes as no surprise that triangles occupy the lion's share of Euclid's similarity arguments.

One such result is the important Proposition VI.8.

PROPOSITION VI.8 If in a right-angled triangle a perpendicular be drawn from the right angle to the base, the triangles adjoining the perpendicular are similar both to the whole and to one another.

PROOF In light of the earlier propositions of Book VI, this was now quite simple. In Figure 3.6, $\triangle BAC$ and $\triangle BDA$ both contain right angles, at $\angle BAC$ and $\angle BDA$, respectively, and both share $\angle 1$. By I.32, their third angles are likewise equal. Similarity, and consequently the proportionality of the sides, then followed from VI.4. The similarity of triangles *BAC* and *ADC*, and that of the two smaller triangles *BDA* and *ADC* were proved in like manner.

Q.E.D.

With the thirty-third and final proposition of Book VI, Euclid had essentially completed his development of plane geometry. Yet, as often surprises those who regard the *Elements* as merely a geometry text, he still had seven books to go. The topic that next fell under his scrutiny proved to be a gold mine for later mathematicians, with a history as rich

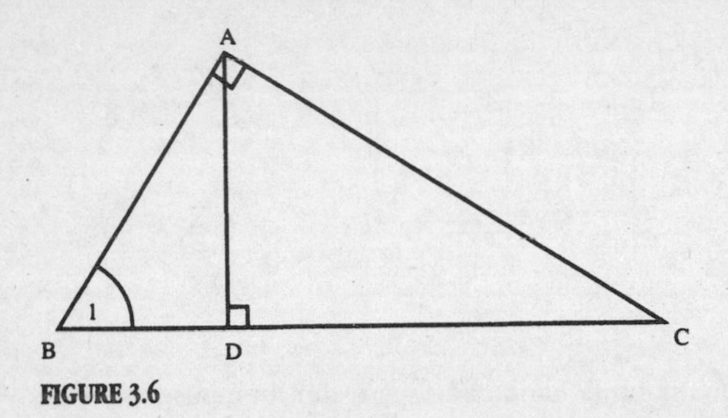

FIGURE 3.6

and glorious as any branch of the subject. It was the theory of numbers, and it is here that the next great theorem will be found.

Number Theory in Euclid

At first glance, one is tempted to dismiss the study of whole numbers as utterly trivial. After all, there seems to be little challenge in such problems as $1 + 1 = 2$ or $2 + 1 = 3$, especially when compared to the intricacies of plane geometry. But any sense of the superficiality of number theory must soon be jettisoned, for this area of mathematics has generated provocative and puzzling questions that have challenged generations of mathematicians. And it is in Books VII through IX of Euclid's *Elements* that we find our oldest significant development of the subject.

Book VII began with a list of 22 new definitions specific to the properties of whole numbers. For instance, Euclid defined an *even* number to be one that is divisible into two equal parts and an *odd* number to be one that is not. A critical definition was that of a *prime* number, that is, a number greater than 1 that is divisible by (Euclid said, "measured by") only 1 and itself. For instance 2, 3, 5, 7, and 11 are primes. Non-prime numbers greater than 1 are called *composite;* each must have a divisor other than one and itself. The first few of these are 4, 6, 8, 9, 10, and 12. The number 1, by the way, is neither prime nor composite.

Further, Euclid defined a *perfect* number to be one which is the sum of its "parts"—that is, its proper divisors. Thus, the number 6 is perfect since its proper divisors are 1, 2, and 3 (we exclude 6 as a divisor of itself, since we want only *proper* divisors), and clearly $1 + 2 + 3 = 6$. The next perfect number is 28, for the sum of its proper divisors is $1 + 2 + 4 + 7 + 14 = 28$. On the other hand, a number like 15 fails to pass muster, since the sum of its proper divisors is $1 + 3 + 5 = 9 \neq 15$, an obvious imperfection. Perfect numbers have long held a special fascination for numerologists and other pseudoscientists, who never fail to find 6s and 28s turning up in the most important and suggestive places. Euclid, fortunately, confined his investigations of perfect numbers to their *mathematical* properties.

Having defined his terms, Euclid got off to a fast start in the first two propositions of Book VII by establishing what has since come to be called "Euclid's algorithm." This is a sure-fire technique for finding the greatest of all the common divisors of two whole numbers. For a brief illustration of the algorithm in action, determine the greatest common divisor of the numbers 1387 and 3796.

Begin by dividing the smaller into the larger and keeping track of the remainder. In this case,

$$3796 = (1387 \times 2) + 1022$$

Next divide the first remainder 1022 into the first divisor 1387 to get

$$1387 = (1022 \times 1) + 365$$

Then repeat the process, this time dividing the second remainder 365 into 1022:

$$1022 = (365 \times 2) + 292 \quad \text{and then}$$

$$365 = (292 \times 1) + 73 \quad \text{and finally}$$

$$292 = (73 \times 4)$$

at which point the remainder is 0.

Upon reaching a zero remainder, Euclid asserted that the *previous* remainder—in our example, 73—is the greatest common divisor of our two original numbers, 1387 and 3796, and he gave a nice proof of this fact. Note that his procedure must eventually terminate, since the remainders—1022, 365, 292, 73—are getting smaller and smaller. When dealing with whole numbers, the process certainly cannot go on forever; in fact, noting that the first remainder was 1022, we can say with absolute certainty that it could at most take 1023 steps before the remainder was whittled away to 0 (of course, it actually took only five steps to do the job).

It is clear that Euclid's algorithm has concrete applications and is entirely automatic. It requires no particular insight or ingenuity to determine the greatest common divisor of a pair of numbers; indeed, a computer can easily be programmed to carry out the process. Less clear, perhaps, is that Euclid's algorithm is also of immense theoretical importance in number theory, where it remains a cornerstone of the subject.

Euclid continued his development of number theory throughout Book VII. Along the way, he came to the crucial Proposition VII.30, which proved that, if a prime number p divides evenly into the product of numbers a and b, then the prime p must divide evenly into one (or both) of a and b, separately. For instance, the prime 17 divides evenly into $2720 = 34 \times 80$, and, sure enough, 17 divides into the first factor, 34. On the other hand, the composite number 12 divides evenly into $48 = 8 \times 6$, but 12 fails to divide into either of the factors 8 or 6 separately. The trouble, of course, is that 12 is not a prime.

Proposition VII.31 will be of importance for the upcoming great theorem. Euclid's proof, identical to that found in modern number theory texts, proceeded as follows.

PROPOSITION VII.31 Any composite number is measured by [that is, divisible by] some prime number.

PROOF Let A be a composite number. By definition of "composite," there must be some smaller number B dividing evenly into A, where $1 < B < A$. Now either B is prime or it is not. If B is prime, then the original number A indeed has a prime divisor, as claimed. Otherwise, B is not prime and so has a divisor, say C, with $1 < C < B$. If C is prime, we are done, for C divides evenly into B, and B divides evenly into A, so the prime C itself divides evenly into A. But what if C is composite? Then, it must have a proper divisor, D, and we continue our quest.

In the worst case, we would keep getting nonprime divisors of descending size:

$$A > B > C > D > \ldots > 1$$

But all of these are positive whole numbers. As Euclid correctly observed, we must reach a point at which the divisor we find is a prime, for ". . . if [a prime divisor] is not found, an infinite series of numbers will measure the number A, each of which is less than the other: which is impossible in numbers." The impossibility, of course, arises simply because a decreasing chain of positive whole numbers can have only finitely many links. We thus can rest assured that the process concludes. The final number in the chain must be a prime as well as a divisor of all numbers before it, and in particular a divisor of the original number A.

Q.E.D.

Both here and in his algorithm, Euclid exploited the key idea that, beginning with any whole number n, a decreasing sequence of positive whole numbers less than *n* must be finite. This is certainly not true if we expand our horizons to fractions, for the descending sequence of positive fractions

$$\frac{1}{2} > \frac{1}{3} > \frac{1}{4} > \frac{1}{5} > \ldots$$

goes on forever. Alternately, if we allow negative whole numbers, a decreasing chain can be endless, as in:

$$32 > 22 > 12 > 2 > -8 > -18 > -28 > \ldots$$

But when we restrict our attention to the *positive integers*, as Euclid did, then such descending sequences must terminate in a finite number of steps, and herein lay the secret of many of his number theoretic deductions.

When Euclid finished the last proof of Book VII, he launched directly into Book VIII without the slightest hitch. In fact, there is no very good reason why his three number theoretic books could not have been merged into a single, albeit quite long, book of the *Elements*. Eventually he arrived at the important Proposition IX.14.

PROPOSITION IX.14 If a number be the least that is measured by prime numbers, it will not be measured by any other prime number except those originally measuring it.

Translated into modern terms, the proposition asserted that a number can be factored into the product of primes in only one fashion. That is, once we have factored ("measured") a number into primes, it is pointless to try to find a factorization into a different collection of primes, for no other primes can measure the original number. Today we call this the "unique factorization theorem" or alternately the "fundamental theorem of arithmetic." The latter name indicates its central role in number theory or, as the subject is sometimes called, the "higher arithmetic."

The unique factorization theorem is used, for instance, in the following little problem. Suppose we begin with the number 8 and decide to take successively higher powers: $8^2 = 64$, $8^3 = 512$, $8^4 = 4096$, $8^5 =$ 32,786, and so forth. We intend to continue until we come to a number that ends with the digit "0." The question is, will it take a hundred steps, or a thousand, or a million to yield such a number?

A quick application of the unique factorization theorem reveals that our quest is utterly hopeless. For, suppose the process eventually yielded a number *N* ending in the digit 0. On the one hand, since *N* was obtained by multiplying a sequence of 8s, we can factor it into a long string of 2s, since $8 = 2 \times 2 \times 2$. However, if *N* ends with digit 0, then it is divisible by 10 and hence by the prime 5. But this is contradictory since, as Euclid proved in IX.14, if N is factored into a string of 2s, then no other prime—in particular the prime 5—can divide evenly into N. In short, even if we keep multiplying 8s together for a million centuries, we can never get a number that ends in 0.

It should be clear from many of the preceding propositions that primes play a central role in the theory of numbers. In particular, since any number beyond 1 is either itself a prime or can be written as the product of primes in a unique way, we can rightly regard the primes as the building blocks of the whole numbers. In this sense the primes from mathematics correspond to the atoms from elementary chemistry and deserve the same kind of intense scrutiny.

Well before Euclid's day, mathematicians had listed the first primes, looking for patterns or other clues to their distribution. Just for reference, the three dozen smallest primes are

2, 3, 5, 7, 11, 13, 17, 19, 23, 29, 31, 37, 41, 43,
47, 53, 59, 61, 67, 71, 73, 79, 83, 89, 97, 101,
103, 107, 109, 113, 127, 131, 137, 139, 149, 151

No particular patterns are immediately evident, except for the obvious one that all primes except 2 are odd numbers (since all larger even numbers have a factor of 2). But a closer look suggests that the primes seem to be "spreading out" or getting scarcer as the numbers grow larger. For instance, there are eight primes between 2 and 20, but only four between 102 and 120. Further, note the gap of 13 consecutive composite numbers between 113 and 127. There is no such long gap among the first 100 numbers.

It is fairly easy to devise an explanation for this apparent "thinning" of the primes. Clearly, when we look at small numbers—those in the teens or twenties—there are fewer *possible* factors since fewer numbers are less than these. By the time we consider larger numbers—such as numbers in the hundreds, or thousands, or millions—there is a multitude of smaller numbers to serve as potential divisors. To be prime, a number must have no smaller factors, and this is considerably less likely for a large number with so many possible divisors below it.

In fact, if we track the primes far enough, we can find huge gaps among them. For instance, of the hundred numbers between 2101 and 2200, only ten are prime, and of the hundred between 10,000,001 and 10,000,100, only two are prime. It would probably have occurred to the Greeks, as it occurs to students today, that the primes may eventually run out. That is, the primes might finally become so scarce as to disappear altogether, with all subsequent numbers composite.

If some evidence seemed to point in this direction, it was not enough to sway Euclid. On the contrary, in Proposition IX.20 he proved that, the thinning notwithstanding, no finite collection of primes could possibly include all the primes there are. His is often called a proof of the "infinitude of primes," for indeed he established that the set of all prime numbers is not finite. Euclid's argument here is a genuine classic, a great theorem if ever there was one. In fact, it is sometimes cited as the finest example of a mathematical theorem that is at once simple, elegant, and extremely profound. The twentieth century British mathematician G. H. Hardy (1877–1947), in his wonderful monograph *A Mathematician's Apology*, called Euclid's proof ". . . as fresh and significant as when it was discovered—two thousand years have not written a wrinkle on [it]."

Great Theorem: The Infinitude of Primes

We have now seen all but one of the ingredients Euclid needed to construct his ingenious proof. What is missing is the very simple observation that if a whole number G divides evenly into both N and M, where $N > M$, then G surely divides evenly into their difference $N - M$. This is easily seen, since G dividing into N means that $N = G \times A$ for some whole number A; and G dividing into M says that $M = G \times B$ for some whole number B. Thus $N - M = G \times A - G \times B = G \times (A - B)$, and since $A - B$ is itself a whole number, G clearly divides evenly into $N - M$. For instance, this says that the difference of two multiples of 5 is itself a multiple of 5; the difference of two multiples of 8 is itself a multiple of 8; and so on.

With this obvious principle behind us, we are ready to attack Euclid's classic result.

PROPOSITION IX.20 Prime numbers are more than any assigned multitude of prime numbers.

Again, Euclid's peculiar terminology partially obscures the proposition's meaning. What he was saying was, given any finite collection of prime numbers—that is, any "assigned multitude"—it is possible to find a prime not contained in this collection. In short, no finite set of primes could possibly exhaust all the primes.

PROOF Euclid began with a finite batch of primes, say $A, B, C, \ldots, D$. His goal was to find a prime number different from all of these. As a first step toward this end, he formed the number $N = (A \times B \times C \times \ldots \times D) + 1$. This number, being one more than the product of all the primes in his initial list, was clearly larger than any of those primes individually. Like any number greater than 1, N is itself either prime or composite, and each of these cases required a separate examination.

CASE 1 Suppose N is prime.

Since it is larger than $A, B, C, \ldots, D$, then N itself is a new prime not included among the originals, and the proof in this case is completed.

CASE 2 What if N is composite?

By Proposition VII.31, N must have a prime divisor, say G. Euclid then asserted—and here lay the heart of his reasoning—that G could not be among the original list of primes in his "assigned multitude." Suppose, for the sake of argument, that $G = A$. Then G surely divides evenly into the product $A \times B \times C \times \ldots \times D$, while (as we have assumed in Case 2), G simultaneously divides evenly into N. Hence, G must also divide evenly into the difference of these numbers, that is, into

$$N - (A \times B \times C \times \ldots \times D)$$
$$= (A \times B \times C \times \ldots \times D) + 1 - (A \times B \times C \times \ldots \times D) = 1$$

But this is impossible, for the prime number G must be at least as big as 2, and no such number can divide evenly into 1. But the same situation exists if we imagined that $G = B$, or $G = C$, and so on. Thus, as Euclid claimed, the prime G is *not* included among $A, B, C, \ldots, D$.

Consequently, whether or not N is prime, a new prime can be found. Hence, any finite collection of primes can always be supplemented by yet another.

Q.E.D.

The thrust of Euclid's argument can be illustrated by considering two specific numerical examples. Suppose, for instance, that our original "assigned multitude" of primes was the set $\{2, 3, 5\}$. Then the number $N = (2 \times 3 \times 5) + 1 = 31$ is itself a prime. Since 31 is clearly larger than the three primes 2, 3, and 5 with which we started, it is a new prime not contained in our original collection. This is exactly the situation covered in the first case.

On the other hand, we might begin with primes $\{3, 5, 7\}$ so that $N = (3 \times 5 \times 7) + 1 = 106$. Now, while 106 is surely bigger than 3, 5, or 7, it is not itself a prime. But, as the second case revealed, 106 *must* have a prime divisor—in this case $106 = 2 \times 53$, and both 2 and 53 serve as new primes not included in $\{3, 5, 7\}$. So, even when N is composite, we can augment our finite list with yet another prime.

This proof will always remain a mathematical classic. But Euclid was not quite done with his number-theoretic investigations. After proving a few rather uninspiring results, such as the fact that the difference of two odd numbers is even, he concluded Book IX with a proposition about perfect numbers. He had defined these at the outset of Book VII but then seemed to have forgotten about them completely. At last, they made their appearance.

PROPOSITION IX.36 If as many numbers as we please beginning from a unit be set out continuously in double proportion, until the sum of all becomes prime, and if the sum multiplied into the last make some number, the product will be perfect.

With the advantages of modern notation, we can express more precisely what Euclid meant: if we begin with 1 and add to it successively higher powers of 2 so the resulting sum $1 + 2 + 4 + 8 + \ldots + 2^n$ is a prime number, then the number $N = 2^n(1 + 2 + 4 + 8 + \ldots + 2^n)$—formed by multiplying the sum $1 + 2 + 4 + 8 + \ldots + 2^n$ by its "last" summand 2^n—must be perfect.

We shall not look at Euclid's proof of this result but shall instead consider a specific example or two. For instance, $1 + 2 + 4 = 7$ is prime, and so, according to Euclid's theorem, the number $N = 4 \times 7 = 28$ is perfect. Of course, we have already verified this. Or, consider $1 + 2 + 4 + 8 + 16 = 31$, a prime. Then $N = 16 \times 31 = 496$ should be perfect. To see that it is, we list the proper divisors of 496—namely, 1, 2, 4, 8, 16, 31, 62, 124 and 248—and add them to get 496, as promised.

Note in passing that numbers of the form $1 + 2 + 4 + \ldots + 2^n$ need not be prime at all. For instance, $1 + 2 + 4 + 8 = 15$ or $1 + 2 + 4 + 8 + 16 + 32 = 63$ are composite. Euclid's perfect number theorem applied only to those *special* cases where this sum indeed turns out to be a prime. Such primes, like 7 and 31, are today called "Mersenne primes," in honor of the French Father Marin Mersenne (1588–1648), who discussed them in a 1644 paper. Because of their link to perfect numbers, Mersenne primes hold a particular fascination for number theorists even to the present day.

In any case, with his proof of Proposition IX.36, Euclid had given a very nice recipe for generating perfect numbers. We shall return to this topic and discuss its current status in the Epilogue.

The Final Books of the *Elements*

In Books VII through IX, Euclid had proved a total of 102 propositions about whole numbers. Then, abruptly, he moved in a different direction in Book X, the longest and, in the opinion of many, the most mathematically sophisticated of the 13. In 115 propositions of Book X, Euclid thoroughly addressed the issue of incommensurable magnitudes, topics that we today would translate into expressions involving square roots of real numbers. Many of these subtle results are technically intricate, involving concepts that need to be defined and examined with care. As an example, consider:

PROPOSITION X.96 If an area be contained by a rational straight line and a sixth apotome, the side of the area is a straight line which produces with a medial area a medial whole.

Obviously it would take some work to sort out the meaning of such terms as "apotome" and "medial" in order to make any sense of Euclid's statement, let alone to understand his subsequent proof. For modern readers, many of these propositions appear obsolete, concerned as they are with topics that are now easily handled within the systems of rational and irrational numbers.

Books XI through XIII cover the fundamentals of solid, or three-dimensional, geometry. The eleventh book, for instance, has 39 propositions examining the solid geometry of intersecting planes, plane angles, and so on. One of its major results was Proposition XI.21, in which Euclid considered a "solid angle"—that is, a three-dimensional angle, such as the apex of a pyramid, formed by three or more plane angles meeting in a point. Euclid proved that the sum of the plane angles converging at this point is less than four right angles. Although we shall not examine Euclid's clever proof, we can easily believe its validity by recognizing that a solid angle containing four right angles—in modern terms 360°—in its plane angles would be "squashed flat" into a plane surface and thus be no angle at all. Proposition XI.21 will figure prominently in the very last result of the *Elements*' very last book.

If Book XI dealt with the elementary propositions of solid geometry, Book XII probed much deeper. Here Euclid employed Eudoxus' method of exhaustion to address such issues as the volume of a cone.

PROPOSITION XII.10 Any cone is a third part of the cylinder which has the same base with it and equal height (Figure 3.7).

We today would express this result as a formula. We know that a cylinder with radius r and height h has volume $\pi r^2 h$, so Euclid was saying that the volume of a cone is $\frac{1}{3}\pi r^2 h$. His wonderful argument was a testament not only to Euclid's expository skills but to the great Eudoxus, who is credited with its initial discovery. Many years later, Archimedes would attribute this proposition to Eudoxus and observe that

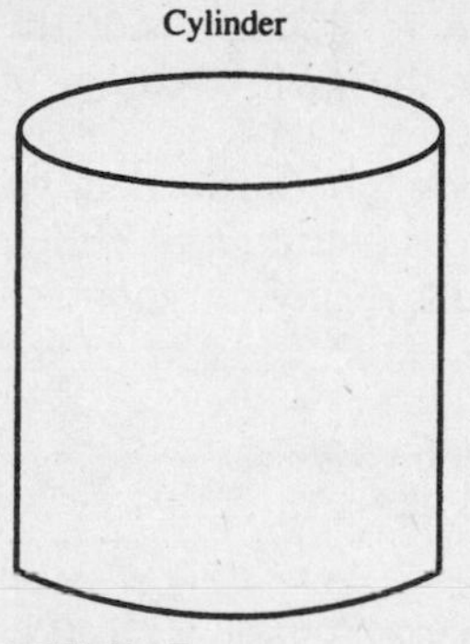

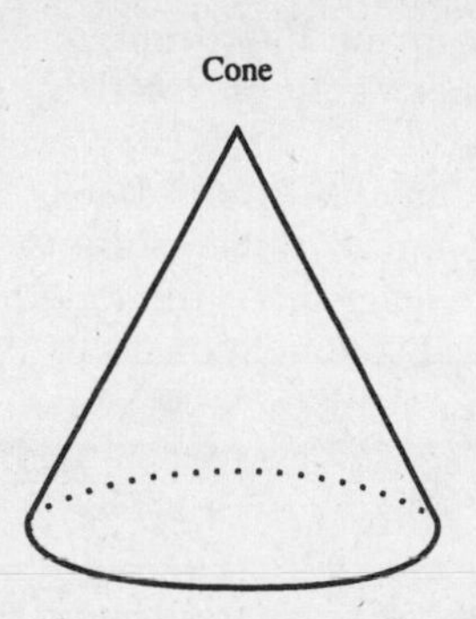

FIGURE 3.7

> . . . though these properties were naturally inherent in the figures all along, yet they were in fact unknown to all the many able geometers who lived before Eudoxus, and had not been observed by anyone.

Book XII contained two other highly significant theorems that deserve mention. The first of these, Proposition XII.2, was somewhat surprisingly about the circle, a *plane* figure.

PROPOSITION XII.2 Circles are to one another as the squares on their diameters.

We encountered this result earlier, when it was used in Hippocrates' quadrature of the lune. As noted then, the proposition provides a means of comparing two circular areas rather than of determining the area of a single circle by knowing its diameter or radius.

Consider XII.2 in a slightly different light. For a pair of circles, one having area A_1 and diameter D_1, and the other having area A_2 and diameter D_2, we conclude that

$$\frac{A_1}{A_2} = \frac{D_1^2}{D_2^2} \quad \text{or equivalently} \quad \frac{A_1}{D_1^2} = \frac{A_2}{D_2^2}$$

This tells us that the ratio of a circle's area to the square of its diameter is always the same—mathematicians would say the ratio is "constant"—regardless of what circle we are considering. This was a highly significant fact. Yet Euclid failed to give a numerical estimate for this constant or to relate it to other important constants one encounters in the study of circles. In short, for all its impressive power, Proposition XII.2 had much room for improvement. We shall return to it later, with the improvement ably provided by Archimedes, as the great theorem of Chapter 4.

In a similar vein, the last proposition of Book XII established, via exhaustion, that "Spheres are to one another in triplicate ratio of their respective diameters." In modern terminology, this relativistic approach to spherical volume reduces to

$$\frac{V_1}{D_1^3} = \frac{V_2}{D_2^3}$$

(Note that taking a "triplicate ratio" was the Greek expression for what we would call cubing.) Here, there emerged another key constant—this time the ratio of spherical volume to the cube of the diameter—but again Euclid gave no hint of what the constant might be. The reader should not be surprised to learn that Archimedes would resolve this one too, in his undisputed masterpiece *On the Sphere and the Cylinder* of 225 B.C.

At last, we come to the thirteenth and final book of Euclid's *Elements*. Its 18 propositions consider the so-called "regular solids" of three-dimensional geometry and the beautiful relationships among them. A regular solid is one all of whose plane faces are congruent, regular polygons. The most familiar of these is the cube, a six-faced solid, each of whose faces is a regular quadrilateral—that is, a square. To the Greeks, the regular solids represented the epitome of beauty and symmetry in three dimensions, and an understanding of these solids would thus have been an obvious priority.

By Euclid's day, five such solids were known—the tetrahedron (a pyramid with equilateral triangles as each of its four faces), the cube, the octahedron (with equilateral triangles as each of its eight faces), the dodecahedron (with regular pentagons as each of its twelve faces), and the most complicated of all, the icosahedron (a 20-faced solid with equilateral triangles as faces).

These aesthetically pleasing solids, shown in Figure 3.8, were featured prominently in Plato's *Timaeus* from around 350 B.C. There Plato considered, among other things, the nature of the four "elements" thought to compose the world—fire, air, water, and earth. It was clear, said Plato, that these four elements are bodies and that all bodies are solids. Since the universe could only have been created out of perfect bodies, it seemed evident (to Plato, at any rate) that fire, air, water, and earth must be in the shape of regular solids. It only remained to determine which element had which shape.

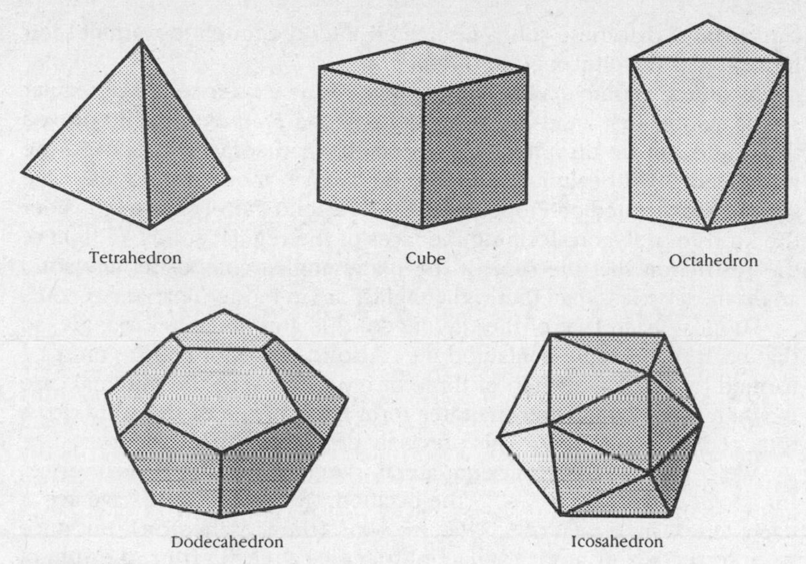

FIGURE 3.8

Plato marshaled his evidence. In the process, he came up with such amusing pseudomathematical statements as ". . . air is to water as water is to earth." His final assignment was as follows.

Fire is in the shape of the tetrahedron, for fire is the smallest, lightest, most mobile, and sharpest of the elements and the tetrahedron fits this description. Earth, said Plato, must be in the shape of the cube, the most stable of the five solids, while water, the most mobile and fluid of the elements, must have as its shape, or "seed," the icosahedron, the solid most nearly spherical and thus most likely to roll easily. Air, somewhat intermediate in size, weight, and fluidity, is composed of octahedrons. "We must," said Plato, "think of the individual units of all four bodies as being far too small to be visible, only becoming visible when massed together in large numbers."

Somewhat embarrassingly, this put Plato in the unenviable situation of having run out of elements but still having a regular solid, the dodecahedron, left over. He lamely said this was the shape ". . . which the god used for arranging the constellations on the whole heaven." In other words, the dodecahedron somehow represented the shape of the universe. Because of this fanciful if not utterly bizarre theory in *Timaeus*, the regular solids have since been called the "Platonic solids." Recalling that Euclid is thought to have studied at Plato's Athenian Academy, one can surmise that these solids fascinated Euclid enough to warrant their inclusion as the climax of the *Elements*.

As noted, geometers had long known of the existence of five regular solids. As the 465th and last proposition of the *Elements*, Euclid proved that there can be no others, that geometry had somehow dictated the number of such beautiful figures to be five, no more and no less. The simple proof relied on Proposition XI.21. Euclid merely had to consider the kinds of polygons forming the faces of the regular solids, in light of the restriction that the sum of the plane angles composing any solid angle must be less than four right angles, or (in modern parlance) 360°.

Suppose each face of the regular solid is an equilateral triangle, so that each plane angle contained 60°. A solid angle, of course, must be formed by the intersection of three or more faces, so the minimal case is when three equilateral triangles form each vertex of the solid, for a total of $3 \times 60° = 180°$. This precisely describes the tetrahedron.

We could also have four equilateral triangles meeting at each vertex, for a total of $4 \times 60° = 240°$ (the octahedron); or we could have five at each vertex, for a total of $5 \times 60° = 300°$ (the icosahedron). But once we intersect six or more equilateral triangles at each vertex, the sum of the plane angles would be at least $6 \times 60° = 360°$, and this violates Proposition XI.21. Thus, there are no other regular solids with equilateral triangles as faces.

What about those whose faces are squares? Each angle of the square is, of course, 90°, so three squares could intersect for a solid angle totaling $3 \times 90° = 270°$; this is the cube. But if four or more squares formed a solid angle, the degree sum would again be at least $4 \times 90° = 360°$, an impossibility. As a consequence, no other regular solids have square faces.

Alternately, the faces may be regular pentagons. Since each interior angle of such a pentagon contains 108°, there can be three ($3 \times 108° = 324° < 360°$) but no more, forming a solid angle. The regular solid thus described is the dodecahedron.

If we try to create one of these solids having as faces regular hexagons, heptagons, octagons, and so forth, then each plane angle will contain at least 120°, so even by putting the minimum of three at each solid angle, we still equal or exceed the limit of 360°. In Euclid's words, "Neither again will a solid angle be contained by other polygonal figures [beyond the regular pentagon] by reason of the same absurdity."

To summarize, Euclid had shown that there can be no more than five regular solids—three with equilateral triangles as faces and one each with squares and regular pentagons as faces. No amount of effort or ingenuity will produce any more of these remarkable figures.

With this, the *Elements* came to an end. It was, and has remained for

2300 years, an unsurpassed mathematical document. As with all great masterpieces, it can be read and reread and yet still provide new insights into the genius of its creator. Even today, these ancient writings can be a source of endless enjoyment for those who take pleasure from the craftsmanship and ingenuity of an elegant mathematical argument. We can do no better than to quote again Sir Thomas Heath, who put it simply, directly, and accurately: The *Elements* ". . . is and will doubtless remain the greatest mathematical textbook of all times."

Epilogue

The great theorem of this chapter involved number theory, so this may be a good time to look ahead at some of the important and often troublesome problems that came to dominate this fascinating branch of mathematics. One of the genuine attractions of number theory is that conjectures simple enough to be understood by elementary school students nonetheless have been immune to the efforts of generations of the world's best mathematicians. It seems an especially perverse feature of this corner of mathematics.

For instance, mathematicians have been intrigued by the phenomenon of "twin primes"—that is, consecutive primes that differ by 2. Examples are 3 and 5, or 11 and 13, or 101 and 103. Like the primes themselves, the prime twins seem to be thinning out as we examine larger and larger tables of numbers. This suggests the obvious question, "Are there only finitely many twin prime pairs?"

It is a simple question. Further, its similarity to the question Euclid settled 2300 years ago in Proposition IX.20 suggests an easy solution. Yet to this day no mathematician knows the answer. It may be, as most mathematicians suspect, that there is no limit to the number of twin primes, but so far no one has been able to prove this. Or it is possible that after a certain point we come to the very largest pair of twins, but no one has proved this either. The situation is, in short, as perplexing as it would have been to Euclid himself. This is a sobering and humbling thought.

Number theory holds other tantalizing but unresolved puzzles. We have mentioned Euclid's proof that if the term in parentheses is a prime, then any number having the form

$$2^n(1 + 2 + 4 + 8 + \dots + 2^n)$$

is perfect. He did not, however, claim that these were the *only* perfect numbers (although neither did he claim they were not). Consequently, mathematicians have tried to find perfect numbers other than those covered by Euclid's formula.

To date, the quest has been thoroughly unsuccessful. In a posthumous paper, the eighteenth century mathematician Leonhard Euler proved that any *even* perfect number must have the form specified by Euclid. That is, if N is an even perfect number, then there exists a positive integer n so that

$$N = 2^n(1 + 2 + 4 + 8 + \ldots + 2^n)$$

where the expression in parentheses must be a (Mersenne) prime.

Between them, Euclid and Euler had completely solved the riddle of even perfect numbers. All that remained was to determine the form of the odd perfect numbers. Unfortunately, no one has ever found one. To this day, whether or not odd perfect numbers exist remains a complete mystery. This is not to say that people have not looked. Centuries of intense theoretical investigations, recently augmented by high-speed computers, have yet to turn up a whole number that is both odd and perfect, although that certainly does not mean there cannot be such numbers of unimaginable size.

Mathematicians are stumped. They neither can find an odd perfect number nor can they prove that such a thing is impossible. This quandary raises an intriguing possibility, however. For, should someone someday furnish a proof that odd perfect numbers do not exist, then all perfect numbers would be even, and as Euler showed, all of these fit Euclid's pattern. In such an eventuality, the great Euclid, in 300 B.C., may have already spotted the pattern that generates the world's entire supply of perfect numbers. That would prove to be quite a remarkable turn of events.

This chapter concludes with one of the most frustrating of all number theoretic questions, the so-called "Goldbach conjecture." It appeared in a 1742 letter of Christian Goldbach (1690–1764), a mathematical enthusiast whose chief claim to fame was that he sent his letter to Euler. Goldbach surmised that any even number greater than or equal to 4 can be written as the *sum* of two primes. Euler tended to agree with Goldbach's assertion but had no inkling as to how to prove it.

As with so many number theoretic puzzles, it is quite easy to check Goldbach's conjecture for small numbers. For instance $4 = 2 + 2$, $28 = 23 + 5$, and $96 = 89 + 7$. The conjecture is particularly tantalizing since it involves such utterly simple concepts. Its only technical terms are "even," "prime," and "sum," and the meanings of these can be conveyed even to young children in just a few minutes. Yet the conjecture has been unresolved since Goldbach mailed his letter two-and-a-half centuries ago.

A peculiar contribution to this problem was made by a Soviet mathematician, L. Schnirelmann. According to historian of mathematics Howard Eves, Schnirelmann proved in 1931 that any even number can be written as the sum of not more than 300,000 primes. Given that Goldbach had conjectured that we only need *two* primes to do the job, Schnirelmann's proof fell substantially short of the mark—299,998 primes short, to be exact.

In a sense, Schnirelmann's 300,000 primes seem to mock the efforts of mathematicians. But they also suggest that, the Euclids and Eulers of history notwithstanding, there still are plenty of great theorems waiting, with their eternal patience, for a proof.

Chapter 4

Archimedes' Determination of Circular Area

(ca. 225 B.C.)

The Life of Archimedes

Two to three generations separated Euclid from the next great mathematician on our agenda, the incomparable Archimedes of Syracuse (287–212 B.C.). By the end of his brilliant career, Archimedes had pushed mathematics well beyond the frontiers of Euclid's day. Indeed, the mathematical world would not see his like again for almost 2000 years.

We are fortunate to have a bit of information about Archimedes' life, although, as with any details coming to us over so many generations, its literal validity can often be challenged. A number of his mathematical works, often prefaced by his own commentaries, have also survived. Taken together, these resources give us a picture of a much revered, somewhat eccentric genius who dominated the mathematical landscape of the classical world.

Archimedes was born at Syracuse on the island of Sicily. His father is thought to have been an astronomer, and as a young boy, Archimedes developed a life-long interest in the study of the heavens. In his youth,

Archimedes also spent some time in Egypt, where he appears to have studied at the great Library of Alexandria. This, of course, had been Euclid's base of operations, and Archimedes would naturally have been trained in the Euclidean tradition, a fact readily apparent in his own mathematical writings.

During his time in the Nile Valley, Archimedes is said to have invented the so-called "Archimedean screw," a device for raising water from a low level to a higher one. Interestingly, this invention remains in use to this day. Its creation testifies to the dual nature of Archimedes' genius: he could concern himself with practical, down-to-earth matters, or could delve into the most abstract, ethereal realms. In spite of Alexandria's obvious appeal to one of his scholarly talents, Archimedes chose to return to his native Syracuse and there, as far as can be determined, spent the rest of his days. Although isolated in Syracuse, he maintained a wide correspondence throughout the Greek world, and particularly with scholars at Alexandria. It is through such correspondence that much Archimedean material has survived.

His awesome mathematical talent was augmented by an ability to devote himself single-mindedly to any problem at hand in extraordinary periods of intense, focused concentration. At such times, the more mundane concerns of life were simply ignored. We learn from Plutarch that Archimedes would

> . . . forget his food and neglect his person, to that degree that when he was occasionally carried by absolute violence to bathe or have his body anointed, he used to trace geometrical figures in the ashes of the fire, and diagrams in the oil on his body, being in a state of entire preoccupation, and, in the truest sense, divine possession with his love and delight in science.

This passage portrays the stereotypically absent-minded mathematician, not to mention one to whom cleanliness was next to irrelevant. Of course, the most famous "absent minded" story concerns the crown of King Hieron of Syracuse. The King, suspicious that his goldsmith had substituted some lesser alloy for the crown's gold, asked Archimedes to determine its true composition. As the story goes, Archimedes wrestled with the problem until one day (during what must have been one of his rare baths) he hit upon the solution. Jumping from the bath, he ran through the streets of Syracuse shouting "Eureka! Eureka!" Unfortunately, so absorbed was he in his wonderful discovery that he forgot to don his toga. What the townspeople thought at seeing their fellow citizen running stark naked in their midst is impossible to say.

This tale may be fictitious, but Archimedes' discovery of the fundamental principles of hydrostatics is pure fact. He left us a treatise titled

On Floating Bodies developing his ideas in this area. Additionally, he advanced the science of optics and did pioneering work in mechanics, as is evident not only in his water pump but in his wonderful understanding of the workings of levers, pulleys, and compound pulleys. Plutarch included the story of a skeptical King Hieron doubting the power of these simple mechanical devices. The King asked for a practical demonstration, and Archimedes obliged in dramatic fashion. He selected one of the King's largest ships

> . . . which could not be drawn out of the dock without great labour and many men; and, loading her with many passengers and a full freight, sitting himself the while far off, with no great endeavour, but only holding the head of the pulley in his hand and drawing the cords by degrees, he drew the ship in a straight line, as smoothly and evenly as if she had been in the sea.

Needless to say, the King was impressed. Perhaps he sensed in this gifted scientist a valuable resource in the event that such engineering talents should be needed for more pressing matters. And indeed they were, when Rome, under the generalship of Marcellus, attacked Syracuse in 212 B.C. In the face of the Roman threat, Archimedes rose to the defense of his homeland by designing an array of weapons of great effectiveness. In the process, he became what can only be called a one-man military-industrial complex.

In what follows, we continue to quote liberally from Plutarch's *Life of Marcellus*, written by the great Roman biographer almost three centuries after the fact. While it was Marcellus about whom Plutarch was ostensibly writing, his admiration for Archimedes was quite evident. These writings provide us with an intriguing—and certainly a very colorful—account of Archimedes in action.

"Marcellus moved with his whole army to Syracuse," Plutarch wrote, "and encamping near the wall, sent ambassadors into the city." When the Syracusans refused to surrender, Marcellus opened his attack on the city walls, both on the land side with his troops and on the ocean side with 60 heavily armed galleys. Marcellus was counting on ". . . the abundance and magnificence of his preparations, and on his own previous glory," but he would prove no match for Archimedes and his diabolical war machines.

According to Plutarch, the Roman legions marched to the city walls, believing themselves to be invincible.

> But when Archimedes began to ply his engines, he at once shot against the land forces all sorts of missile weapons, and immense masses of stone that came down with incredible noise and violence; against which no man could

> stand; for they knocked down those upon whom they fell in heaps, breaking all their ranks and files.

The Roman naval forces fared no better, for

> . . . huge poles thrust out from the walls over the ships sunk some by the great weights which they let down from on high upon them; others they lifted up into the air by an iron hand or beak . . . and, when they had drawn them up by the prow, and set them on end upon the poop, they plunged them to the bottom of the sea; or else the ships, drawn by engines within, and whirled about, were dashed against steep rocks that stood jutting out under the walls, with great destruction of the soldiers that were aboard them.

Such destruction, related Plutarch, was "a dreadful thing to behold," and one is inclined to agree. Under the circumstances, Marcellus thought it prudent to retreat. He withdrew both land and naval forces to regroup. Holding a council of war, the Romans decided upon a night assault, in the expectation that Archimedes' devilish weapons would be useless if the attackers slipped too close to the walls under the cover of darkness. Again, the Romans had an unpleasant surprise. The diligent Archimedes had arranged his devices for just such an eventuality, and no sooner had the Romans crept up close upon the fortifications than "stones came tumbling down perpendicularly upon their heads, and, as it were, the whole wall shot out arrows at them." In response, the terrified Romans again retreated, only to come under attack from Archimedes' longer-range weapons, an attack that "inflicted a great slaughter among them." By this time, the vaunted Roman legions, "seeing that indefinite mischief overwhelmed them from no visible means, began to think they were fighting with the gods."

It is perhaps an understatement to say that Marcellus had a serious morale problem. He demanded of his shaken troops a renewed courage to continue the assault, but the previously invincible Romans wanted no more of it. On the contrary, the soldiers "if they did but see a little rope or a piece of wood from the wall, instantly crying out, that there it was again, Archimedes was about to let fly some engine at them, they turned their backs and fled." Knowing that discretion is the better part of valor, Marcellus chose to abandon the direct assault.

Instead, trying to starve the trapped Syracusans into surrender, the Romans began a long siege of the city. Time passed, with no change in the disposition of forces. Then, during a feast to Diana, the city inhabitants, "given up entirely to wine and sport," became careless about guarding a section of the wall, and the opportunistic Romans saw their chance. Their armies broke through the lightly guarded section and poured into the city in a vicious and destructive mood. Marcellus, surveying the beautiful town, is said to have wept in anticipation of the havoc that his men were sure to wreak. Indeed, history records that the Romans treated Syracuse no less harshly than they would treat Carthage some 66 years later.

But it was the death of Archimedes that brought Marcellus his greatest sorrow, for he had come to respect his gifted antagonist. According to Plutarch,

> . . . as fate would have it, intent upon working out some problem by a diagram, and having fixed his mind alike and his eyes upon the subject of his speculation, [Archimedes] never noticed the incursion of the Romans, nor that the city was taken. In this transport of study and contemplation, a soldier, unexpectedly coming up to him, commanded him to follow to Marcellus; which he declining to do before he had worked out his problem to a demonstration, the soldier, enraged, drew his sword and ran him through.

Thus ended the life of Archimedes. He died, as he had lived, lost in thought about his beloved mathematics. We can regard him either as a martyr to his research or as a victim of his own preoccupied mind. In any case, mathematicians may come and mathematicians may go, but no other has had an end quite like this.

For all of Archimedes' great weapons, for all of his practical inventions, his true love was pure mathematics. His levers and pulleys and catapults were mere trifles compared with the beautiful theorems he discovered. Again, we quote Plutarch:

> Archimedes possessed so high a spirit, so profound a soul, and such treasures of scientific knowledge, that though these inventions had now obtained him the renown of more than human sagacity, he yet would not deign to leave behind him any commentary or writing on such subjects; but, repudiating as sordid and ignoble the whole trade of engineering, and every sort of art that lends itself to mere use and profit, he placed his whole affection and ambition in those purer speculations where there can be no reference to the vulgar needs of life.

It was his mathematics that would be his greatest legacy. In this arena, Archimedes stands unchallenged as the greatest mathematician of antiquity. His results, which survive in a dozen books and fragments, are of the highest quality and show a logical sophistication and polish that is truly astounding. Not surprisingly, he was very familiar with Euclid and proved to be a master of Eudoxus' method of exhaustion; to use Newton's charming phrase, Archimedes surely stood on the shoulders of giants. But past influences, great as they were, cannot adequately explain the amazing advances that Archimedes would bring to the discipline of mathematics.

Great Theorem: The Area of the Circle

Around 225 B.C., Archimedes produced a short treatise titled *Measurement of a Circle*, the first proposition of which gave a penetrating analysis of circular area. Before addressing this classic work, however, we first need to examine what was known about circular areas when Archimedes arrived upon the scene.

Geometers of the time would have known that, regardless of the circle in question, the ratio of the circumference of a circle to its diameter is always the same. In modern terminology, we would say that

$$\frac{C_1}{D_1} = \frac{C_2}{D_2}$$

where C is the circumference and D is the diameter of the circles in Figure 4.1. Put another way, the ratio of a circle's circumference to its diameter is constant, and modern mathematicians *define* π to be this ratio. (Note that the Greeks did not use the symbol in this context.) Thus, the formula

$$\frac{C}{D} = \pi \quad \text{or its equivalent} \quad C = \pi D$$

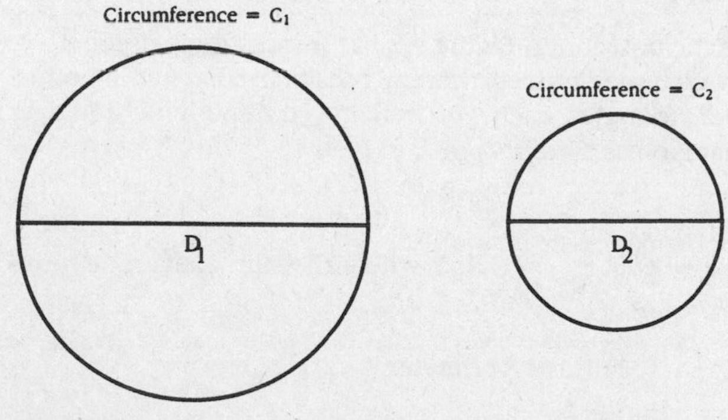

FIGURE 4.1

is nothing more than the definition of the constant π as it arises in the comparison of two lengths—a circle's circumference and its diameter.

But what about circular areas? As we have seen, Proposition XII.2 of the *Elements* established that two circular areas are to each other as the squares on their diameters, and thus the ratio of circular area to the square of the diameter is constant. In modern terms, Euclid had proved that there is some constant k such that

$$\frac{A}{D^2} = k \quad \text{or equivalently} \quad A = kD^2$$

All of this was fine as far as it went. But how do these constants relate to one another? That is, can one find a simple connection between the "one-dimensional" constant π (used in relating circumference to diameter) and the "two-dimensional" constant k (used in relating area to diameter)? Apparently Euclid had found no such connection.

But in his short yet elegant treatise *Measurement of a Circle*, Archimedes proved what amounts to the modern formula for circular area involving π. In doing this, he made the critical link between circumference (and hence π) and circular area. His proof required two fairly direct preliminary results plus a rather sophisticated logical strategy called double *reductio ad absurdum* (reduction to absurdity).

We shall examine these preliminaries first. One concerned the area of a regular polygon with center O, perimeter Q, and apothem h, where the apothem is the length of the line drawn from the polygon's center perpendicular to any of the sides.

THEOREM The area of the regular polygon is $\frac{1}{2}hQ$.

PROOF Suppose the polygon in Figure 4.2 has n sides, each of length b. Draw lines from O to the vertices, thereby breaking it up into a collection of n congruent triangles, each with height h (the apothem) and base b. Since each triangle has area $\frac{1}{2}bh$,

Area (regular polygon)

$$= \tfrac{1}{2}bh + \tfrac{1}{2}bh + \ldots + \tfrac{1}{2}bh, \text{ where the sum contains } n \text{ terms}$$
$$= \tfrac{1}{2}h(b + b + \ldots + b) = \tfrac{1}{2}hQ$$

since $(b + b + \ldots + b)$ is the perimeter.

Q.E.D.

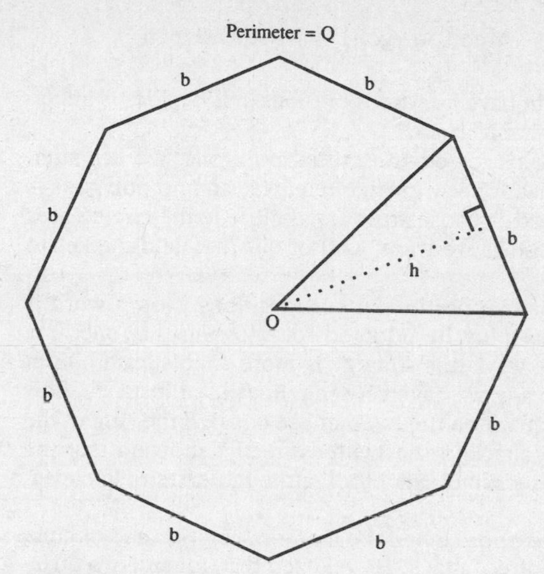

FIGURE 4.2

That was quick enough. Archimedes' other preliminary was also well known in his day, and seems quite self-evident. It says that if we are given a circle, we can inscribe within it a square; Euclid himself gave this construction in Proposition IV.6. The square's area, of course, is less than that of the circle in which it was inscribed. By bisecting each side of the square, we can locate the vertices of a regular octagon inscribed within the circle. Of course, the octagon more nearly approximates the circle's area than the square did. If we again bisected to get a regular 16-gon, it would be closer to the circle in area than the octagon was.

The process can be continued indefinitely. This is, in fact, the essence of Eudoxus' famous method of exhaustion alluded to earlier. Clearly the area of an inscribed polygon never equals that of the circle; there will always be an excess of circle over inscribed polygon regardless of the number of sides of the latter. But—and this was the key to the method of exhaustion—if we have any *preassigned* area, no matter how small, we can construct an inscribed regular polygon for which the difference between the circle's area and the polygon's is less than this preassigned amount. For instance, if we were given a preassigned area of $\frac{1}{100}$ of a square inch, we could come up with a regular inscribed polygon for which

$$\text{Area (circle)} - \text{Area (polygon)} < \tfrac{1}{500} \text{ square inch}$$

That such a polygon might have hundreds or thousands of sides is immaterial; the crucial fact is that it *exists.*

An analogous rule holds for circumscribed polygons. We can summarize both by saying that, for any given circle, we can find polygons—inscribed or circumscribed—whose areas are as close to the circle's area as we want. It is the "close as we want" part of this that held the key to Archimedes' success.

These, then, were his two preliminary propositions. Now a word is needed about the logical ploy he adopted for showing that one area equals another. In some ways this strategy is more sophisticated, or at least more devious, than any we have yet seen. Recall, for instance, how Euclid proved that the square on the hypotenuse equaled the sum of the squares on the legs: he attacked the matter directly, showing that the areas in question were the same. His proof, although extremely clever, was a frontal assault.

But when Archimedes approached the far more complicated circular area, he employed an indirect attack. He realized that, for any two quantities A and B, one and only one of the following cases holds: $A < B$ or $A > B$ or $A = B$. Wanting to prove that $A = B$, Archimedes would first make the assumption that $A < B$ and from this derive a logical contradiction, thereby eliminating the case as a possibility. Next, he would suppose that $A > B$, which again led him to a contradiction. With both of these options eliminated, there remained but one alternative, namely, that A and B are equal.

This was his wonderful, indirect strategy—a "***double reductio ad absurdum***" since it reduced two of the three cases to a contradiction. While this may initially seem a bit roundabout, a little reflection shows it to be quite reasonable; eliminate two of the three possible cases and one is forced to conclude that the third is valid. Certainly no one used double *reductio ad absurdum* more deftly than Archimedes.

With these preliminaries behind us, we can now watch a master at work in the first proposition from *Measurement of a Circle*:

PROPOSITION 1 The area of any circle is equal to a right-angled triangle in which one of the sides about the right angle is equal to the radius, and the other to the circumference, of the circle.

PROOF Archimedes began with two figures (Figure 4.3): a circle having center O, radius r, and circumference C; and a right triangle having base of length C and height of length r. We denote by A the area of the circle

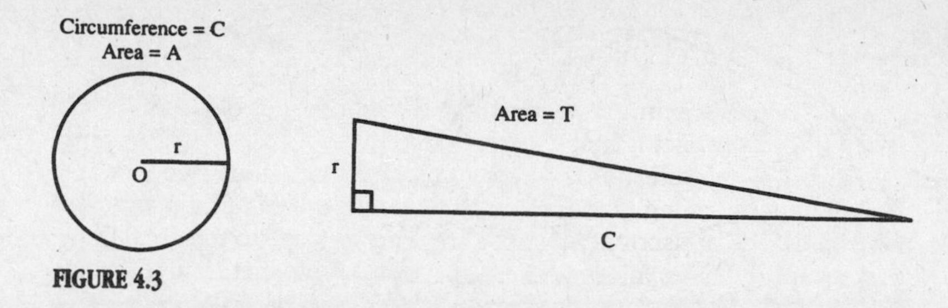

FIGURE 4.3

and by T the area of the triangle. While the former is the object of Archimedes' proof, it is clear that the triangle's area is just $T = \frac{1}{2}rC$.

The proposition claimed simply that $A = T$. To establish this by a double *reductio ad absurdum* proof, Archimedes needed to consider, and eliminate, the other two cases.

CASE 1 Suppose $A > T$.

This asserts that the circular area exceeds that of the triangle by some amount. In other words, the excess $A - T$ is some positive quantity. Archimedes knew that, by *inscribing* a square within his circle and repeatedly bisecting its sides, he could arrive at a regular polygon inscribed within the circle whose area differs from the area of the circle by less than this positive amount $A - T$. That is,

$$A - \text{Area (inscribed polygon)} < A - T$$

Adding the quantity "Area (inscribed polygon) $+ T - A$" to both sides of this inequality yields

$$T < \text{Area (inscribed polygon)}$$

But this is an *inscribed* polygon (Figure 4.4). Thus its perimeter Q is less than the circle's circumference C, and its apothem h is certainly less than the circle's radius r. We conclude that

$$\text{Area (inscribed polygon)} = \tfrac{1}{2}hQ < \tfrac{1}{2}rC = T$$

Here Archimedes had reached the desired contradiction, for he had found both that $T <$ Area (inscribed polygon) and that Area (inscribed polygon) $< T$. There is no logical recourse other than to conclude that Case 1 is impossible; the circle's area cannot be more than the triangle's.

This left him with the second case.

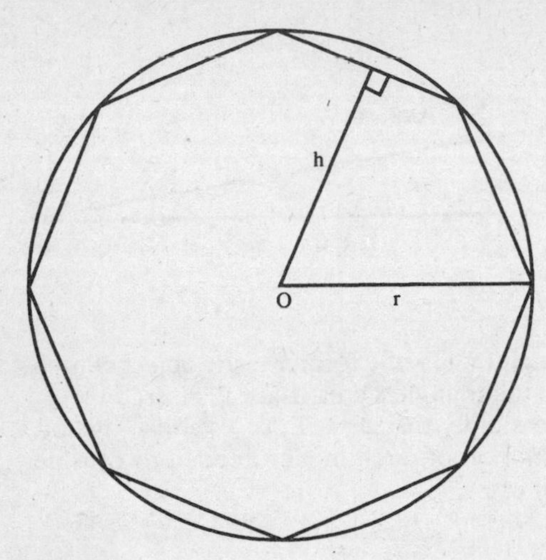

FIGURE 4.4

CASE 2 Suppose $A < T$.

This time Archimedes assumed that the circle's area fell short of the triangle's, so that $T - A$ represented the excess area of the triangle over the circle. We know that we can circumscribe about the circle a regular polygon whose area exceeds the circle's area by less than this amount $T - A$. In other words,

$$\text{Area (circumscribed polygon)} - A < T - A$$

If we simply add A to both sides of the inequality, we conclude that

$$\text{Area (circumscribed polygon)} < T$$

But the circumscribed polygon (Figure 4.5) has its apothem h equal to the circle's radius r, while the polygon's perimeter Q obviously exceeds the circle's circumference C. Thus,

$$\text{Area (circumscribed polygon)} = \tfrac{1}{2}hQ > \tfrac{1}{2}rC = T$$

Again this is a contradiction, since the circumscribed polygon cannot be both less than and greater than the triangle in area. Archimedes con-

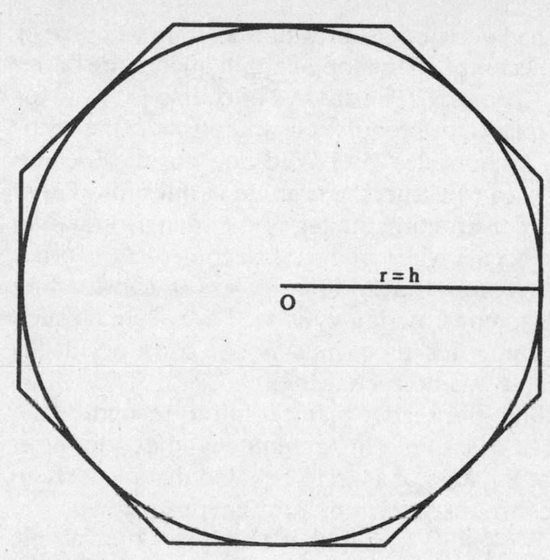

FIGURE 4.5

cluded that Case 2 was likewise impossible; the circle's area cannot be less than the triangle's.

As a consequence, Archimedes could write: "Since then the area of the circle is neither greater nor less than [the area of the triangle], it is equal to it."

Q.E.D.

This was his proof, a little gem from the hand of an indisputably great mathematician. It strikes some people as odd that Archimedes proved the circle's area must equal that of the triangle by showing that it could be neither greater nor less. For those who find his argument a bit too indirect for their taste, a paraphrase of *Hamlet*'s Polonius is offered: "though this be madness, yet there is method of exhaustion in't." One is tempted to wonder how something this short and simple could have been overlooked by Hippocrates or Eudoxus or Euclid. But simplicity is most easily perceived in hindsight. In this regard, we again turn to Plutarch's characterization of Archimedes' mathematics:

> It is not possible to find in all geometry more difficult and intricate questions, or more simple and lucid explanations. Some ascribe this to his natural genius; while others think that incredible effort and toil produced these, to all appearances, easy and unlaboured results. No amount of investigation of yours would succeed in attaining the proof, and yet, once seen, you immediately believe you would have discovered it; by so smooth and so rapid a path he leads you to the conclusion required.

Given that Archimedes had equated the area of the circle with that of a triangle, did he therefore accomplish the long-sought quadrature of the circle that we examined in Chapter 1? The answer of course is "No," for we recall that a successful quadrature requires us to *construct* the rectilinear figure of equal area. Archimedes' proof did not, nor did it claim to, give any inkling as to how to construct the triangle in question. There is, of course, no difficulty in constructing the leg of the triangle equaling the circle's radius; the snag occurs when one tries to construct the other leg equal to the triangle's circumference. Since $C = \pi D$, constructing the circumference amounts to constructing π. As we have seen, no such construction is possible. Archimedes' proof must not be construed as his attempt to square the circle; it was no such thing.

All of this notwithstanding, the reader may yet fail to recognize the familiar formula for the area of a circle in Archimedes' theorem. After all, what he proved was that the area of a circle equaled that of a certain triangle. As we shall see, this was a typical Archimedean device—to relate the area of an unknown figure with that of a simpler, known one. But more was going on than just this. For the triangle in question had as its base the circle's circumference, and this had two crucial implications. First, unlike Euclid, Archimedes had related a circle's area not to that of another circle (basically a "relativistic" approach) but to its own circumference and radius, as reflected in the equivalent triangle. Then, by proving that $A = T = \frac{1}{2}rC$, Archimedes had provided the link between the one-dimensional concept of circumference and the two-dimensional concept of area. Remembering that $C = \pi D = 2\pi r$, we rephrase his theorem as

$$A = \tfrac{1}{2}rC = \tfrac{1}{2}r(2\pi r) = \pi r^2$$

and here emerges one of geometry's most familiar and important formulas.

It is also worth noting that Archimedes' bold proposition easily implied Euclid's relatively tame result that the areas of two circles are in the same ratio as the squares upon their diameters. That is, if we let one circle have area A_1 and diameter D_1 and a second circle have area A_2 and diameter D_2, then Archimedes proved

$$A_1 = \pi r_1^2 = \pi(D_1/2)^2 = \pi D_1^2/4 \quad \text{and} \quad A_2 = \pi r_2^2 = \pi(D_2/2)^2 = \pi D_2^2/4$$

Hence

$$\frac{A_1}{A_2} = \frac{\pi D_1^2/4}{\pi D_2^2/4} = \frac{D_1^2}{D_2^2}$$

which is Euclid's theorem in a nutshell. So, this Archimedean proposition had enough power to imply the Euclidean result as a trivial corollary. Such is the mark of a genuine mathematical advance.

If we look back at the previous discussion, we can now determine the value of the constant k in the "Euclidean" expression $A = kD^2$. For, with Archimedes' discovery at hand, we know that

$$\pi r^2 = A = kD^2 = k(2r)^2 = 4kr^2$$

Hence, $4k = \pi$, and so $k = \pi/4$. In other words, Euclid's "two-dimensional" area constant is just a quarter of π, the "one-dimensional" circumference constant. Thus, his proposition brought the welcome news that we need not calculate two different constants. If we can just determine the value of π from the circumference problem, it would also serve in the formula for circular area.

This latter observation was not lost on Archimedes. In fact, as the third proposition of *Measurement of a Circle*, he derived just such a value.

PROPOSITION 3 The ratio of the circumference of any circle to its diameter is less than $3\frac{1}{7}$ but greater than $3\frac{10}{71}$.

In modern notation, this says: $3\frac{10}{71} < \pi < 3\frac{1}{7}$. With these fractions converted to their decimal equivalents, Archimedes' result becomes 3.140845 . . . $\pi < 3.142857$ . . .; hence, the constant π has been nailed down, to two decimal place accuracy, as 3.14.

That Archimedes came up with this estimate is another sign of his powers. His plan of attack was again to use his ever-helpful inscribed and circumscribed regular polygons, except this time, instead of tracking down their areas, he was concentrating on their perimeters. He began with a regular hexagon inscribed in a circle (Figure 4.6). He knew well that each side of the hexagon equaled the circle's radius, whose length we can call r. Thus,

$$\pi = \frac{\text{circumference of circle}}{\text{diameter of circle}} > \frac{\text{perimeter of hexagon}}{\text{diameter of circle}} = \frac{6r}{2r} = 3$$

Admittedly, this was a very crude estimate for π, but Archimedes had just begun. He next doubled the number of sides of his inscribed polygon, to get a regular dodecagon whose perimeter he had to calculate. This is where he leaves modern mathematicians shaking their heads in wonder, for determining the dodecagon's perimeter required getting a numerical value for the square root of three. With our calculators and

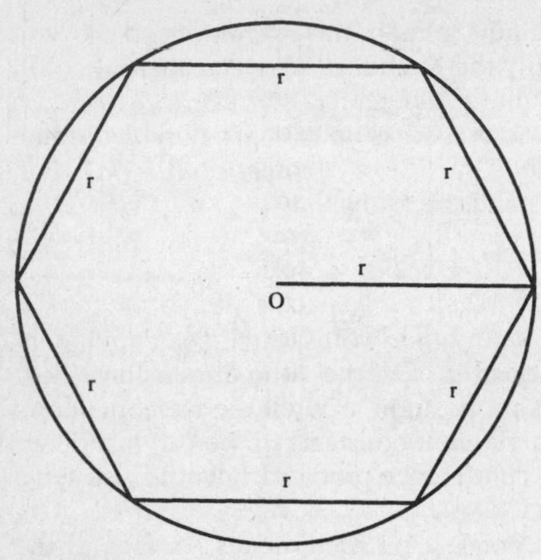

FIGURE 4.6

computers, this strikes us as no real obstacle, but in Archimedes' time, not only were these devices unthinkable, but there was not even a good number system to facilitate such computations. Yet he emerged with the estimate

$$\frac{265}{153} < \sqrt{3} < \frac{1351}{780}$$

which is impressively close.

From there, Archimedes continued, bisecting again to get a regular 24-gon, then a regular 48-gon, and finally a regular 96-gon. At each stage, he needed to approximate sophisticated square roots, yet he never faltered. When he reached the 96-gon, his estimate was

$$\pi = \frac{\text{circumference of circle}}{\text{diameter of circle}}$$

$$> \frac{\text{perimeter of regular 96-gon}}{\text{diameter of circle}} > \frac{6336}{2017\frac{1}{4}} > 3\tfrac{10}{71}$$

As if this were not enough, Archimedes then turned around and made similar estimates for regular *circumscribed* 12-gons, 24-gons, 48-gons, and 96-gons, leading him to his upper bound for π of $3\frac{1}{7}$. Such calculations, in the face of an absolutely terrible numeral system and without easy procedures for estimating the square roots he needed, provide sure evidence of his awesome powers. These computations were the arithmetical counterpart of running the high-hurdles wearing a ball and chain. Yet by marshaling his enormous intellect and perseverance, he succeeded in giving the first scientific estimate of the critical constant π. As indicated in the Epilogue to this chapter, the quest for highly accurate estimates of this number has occupied mathematicians ever since.

As it has come down to us, *Measurement of a Circle* contains only three propositions and covers only a few pages of text. Moreover, the second proposition is out of place and unsatisfactory, undoubtedly the result of bad copying, bad editing, or bad translating years, if not centuries, after Archimedes. On the surface, then, it seems unlikely that such a short work would carry the impact that it does. But considering that in its first proposition, Archimedes proved the famous formula for the area of a circle, and in its last, he gave a remarkable estimate for the number π, there is really no doubt why this little treatise had been held in such high regard by generations of mathematicians. It is not the quantity of pages but the quality of the mathematics, and by this criterion *Measurement of a Circle* stands as a genuine classic.

Archimedes' Masterpiece: *On the Sphere and the Cylinder*

The results just discussed constitute but a fraction of the mathematical legacy of Archimedes. He also wrote about the geometry of spirals and about conoids and spheroids, and he provided a remarkable means of finding the area under a parabola by summing a certain infinite geometric series. This latter topic—finding areas under curves—is now treated in calculus courses, another indication (if one were needed) of how utterly far ahead of his time Archimedes was.

But for all of these accomplishments, his undisputed masterpiece was an extensive, two-volume work titled *On the Sphere and the Cylinder.* Here, with almost superhuman cleverness, he determined volumes and surface areas of spheres and related bodies, thereby achieving for three-dimensional solids what *Measurement of a Circle* had done for two-dimensional figures. It was a stunning triumph, one that Archimedes himself seems to have regarded as the apex of his career.

We should first recall what the Greeks knew about the surface areas and volumes of three-dimensional bodies. As noted in the previous chapter, Euclid had proved that the volumes of two spheres are to each other as the cubes of their diameters; in other words, there exists a "volume constant" m so that

$$\text{Volume (sphere)} = mD^3$$

This was the Euclidean treatment of spherical volume. As to the surface area of a sphere, Euclid was utterly silent. Here again, a successful assault on the problem awaited Archimedes' *On the Sphere and the Cylinder.*

This two-volume work had a familiar ring to it, insofar as it began with a list of definitions and assumptions from which he derived ever more sophisticated theorems. In short, it was cast in the Euclidean mold. Its first proposition was the innocuous: "If a polygon be circumscribed about a circle, the perimeter of the circumscribed polygon is greater than the circumference of the circle." However, Archimedes quickly moved in more sophisticated directions. Throughout, he was (at least to modern tastes) hampered by the lack of a concise algebraic notation. Unable to express his volumes and surface areas by simple formulas, he had to rely on statements such as:

PROPOSITION 13 The surface of any right circular cylinder excluding the bases is equal to a circle whose radius is a mean proportional between the side of the cylinder and the diameter of the base.

At first glance, this looks quite mysterious and unfamiliar, but it is in the phrasing, not the content, that the unfamiliarity lies. Without the benefit of algebra, Archimedes had to express his desired area—in this case that of a lateral surface of a right circular cylinder—as being equal to the area of a known figure—in this case, a circle (Figure 4.7). But which circle? Obviously Archimedes had to specify his equivalent circle, and that is where the statement about mean proportionals came in.

In modern terminology, Archimedes was claiming that

Lateral surface (cylinder of radius r and height h)
= Area (circle of radius x)

where $h/x = x/2r$. From this it follows quickly that $x^2 = 2rh$, and so we get the well-known formula:

$$\text{Lateral surface (cylinder)} = \text{Area (circle)} = \pi x^2 = 2\pi rh$$

Archimedes proceeded through a string of like-sounding propositions as he approached his first major objective, the surface area of the sphere. Space does not allow us to follow him in his reasoning, but we can acknowledge his remarkable ingenuity. In light of our earlier examination of his mathematics, the reader should not be surprised to learn that Archimedes again used the method of exhaustion. That is, he "exhausted" the sphere by approximating it from within and without by

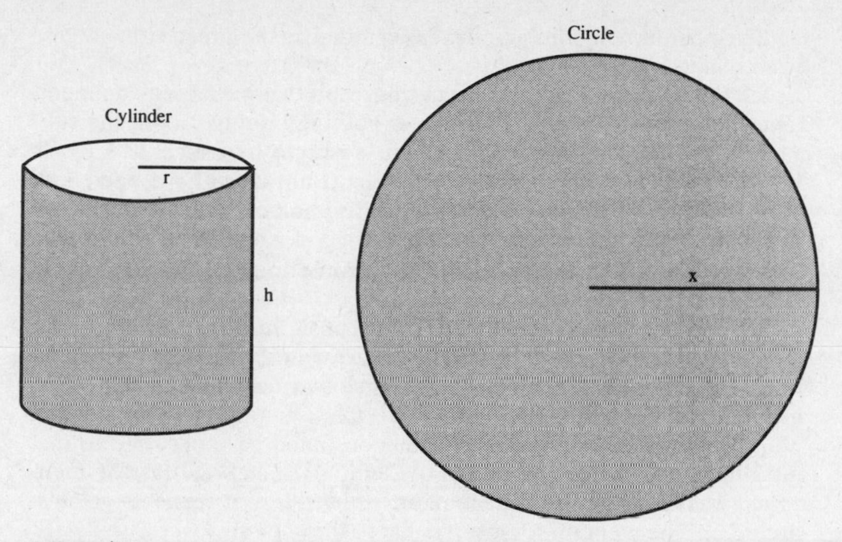

FIGURE 4.7

cones and the frusta of cones, all of whose surface areas he had previously determined. When the dust had settled, he had proved the remarkable

PROPOSITION 33 The surface of any sphere is equal to four times the greatest circle in it.

Archimedes completed the proof with his favorite logical tactic of double *reductio ad absurdum*; that is, he proved it impossible for the spherical surface to be *more* than four times the area of its greatest circle and also proved it impossible for it to be *less* than four times the area of its greatest circle. If we observe that the area of the "greatest circle" of the sphere—that is, the circle through the sphere's "equator"—is just πr^2, then we can translate Archimedes' formulation of this result—"the surface of the sphere is four times the area of its greatest circle"—into the modern-day formula

$$\text{Surface area (sphere)} = 4\pi r^2$$

This is a very sophisticated piece of mathematics. The deftness with which Archimedes handled his concepts, the insights that he brought to bear, seem to anticipate the ideas of modern integral calculus. It is readily apparent why Archimedes is regarded as the greatest mathematician of ancient times.

But there is one other fact about this result that warrants a comment, namely, its utter strangeness. There is nothing intuitive about the substantive fact that the surface of a sphere is *exactly* four times as large as the area of its greatest cross section. Why could it not have been 4.01 times as great? What is so magical about this number "four" to guarantee that if one were to paint the curving surface of a sphere, it would take precisely four times as much paint as it would to paint the great circle through the center?

Archimedes himself addressed this peculiar, intrinsic property of the sphere in his introduction to *On the Sphere and the Cylinder*, which he wrote for a certain "Dositheus," presumably a mathematician at Alexandria to whom Archimedes had sent the treatise. Archimedes noted that ". . . certain theorems not hitherto demonstrated have occurred to me, and I have worked out the proofs of them." First among those he mentioned was ". . . that the surface of any sphere is four times its greatest circle," and he went on to observe that such properties were

> . . . all along naturally inherent in the figures referred to, but remained unknown to those who were before my time engaged in the study of geometry. Having, however, now discovered that the properties are true . . . , I cannot feel any hesitation in setting them side by side both with my former investigations and with those of the theorems of Eudoxus on solids which are held to be most irrefragably established . . .

The comment provides an interesting glimpse of Archimedes' assessment of his work and its place in the development of mathematics. He did not hesitate to include himself alongside the great Eudoxus, for he surely was well aware of the extraordinary nature and quality of his own discoveries. But he also went out of his way to stress that he had not invented or created the fact that $S = 4\pi r^2$. Rather, he had been fortunate enough to *discover* an intrinsic property of spheres, one that had existed since time immemorial even though it had been previously unknown to geometers. To Archimedes, mathematical relationships existed independent of the poor efforts of humans to decipher them. He himself had just been the individual fortunate enough to glimpse these eternal truths.

If *On the Sphere and the Cylinder* had contained nothing but the previous theorem, it would have stood as a classic for all time. But he immediately turned his gaze toward spherical *volume*. After another intricate double *reductio ad absurdum* argument, Archimedes succeeded in establishing

PROPOSITION 34 Any sphere is equal to four times the cone which has its base equal to the greatest circle in the sphere and its height equal to the radius of the sphere.

Note that, again, Archimedes has expressed the volume of the sphere not as a simple algebraic formula but in terms of the volume of a simpler solid, in this case, a cone (Figure 4.8). With just a bit of effort we can convert his verbal statement into its modern equivalent.

That is, let r be the radius of the sphere. Then the "cone which has its base equal to the greatest circle in the sphere and its height equal to the radius of the sphere" is such that

$$\text{Volume (cone)} = \tfrac{1}{3}\pi r^2 h = \tfrac{1}{3}\pi r^2 r = \tfrac{1}{3}\pi r^3$$

But Archimedes' Proposition 34 had proved that the volume of the sphere is four times as great as the volume of one of these cones, and this yields the famous formula

$$\text{Volume (sphere)} = 4\ \text{Volume (cone)} = \tfrac{4}{3}\pi r^3$$

Among its benefits, this result clarifies the link between π and the "volume constant" m that arose from Euclid's Proposition XII.18. Referring to our discussion above, we immediately see that

$$\tfrac{4}{3}\pi r^3 = \text{Volume (sphere)} = mD^3 = m(2r)^3 = 8mr^3$$

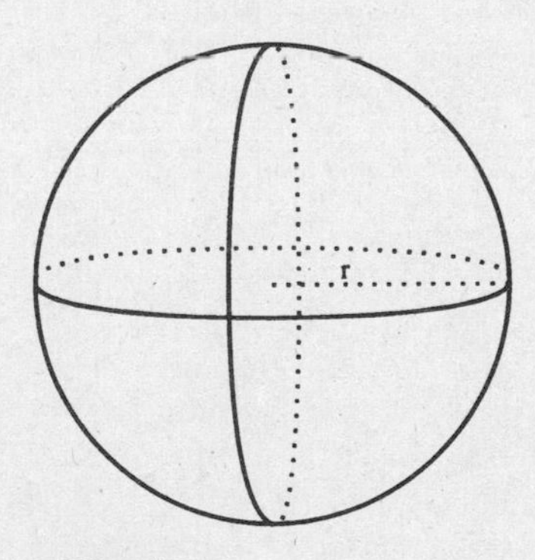

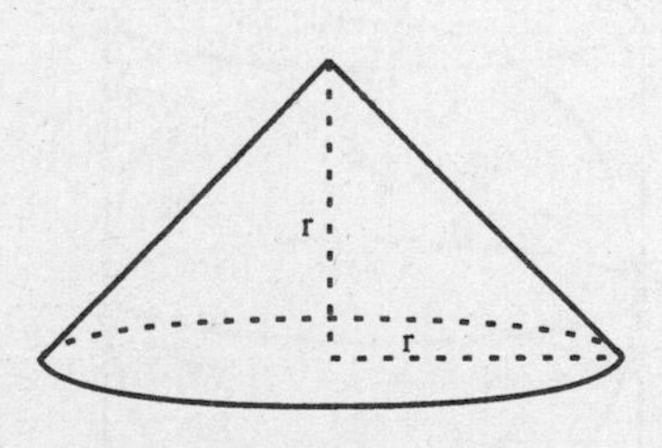

FIGURE 4.8

and a little algebra reveals that $m = \pi/6$. In this fashion, the pre-Archimedean mystery regarding circumferences, circular areas, and spherical volumes was resolved. No longer were three different constants needed to address these three different matters; all three rested upon knowledge of π. Archimedes had exhibited a stunning unity among them.

Immediately upon completing his proofs of Propositions 33 and 34, Archimedes restated his results in a particularly intriguing way. He considered a cylinder circumscribed about the sphere, as shown in Figure 4.9. He then asserted that the cylinder is half again as large as the sphere in *both surface area and volume*! In a certain sense, this was the climax of his whole work. It took his two great results and presented them in a simple fashion, expressing the complicated spherical surface and volume in terms of the correspondingly simpler surface and volume of a related cylinder. This section will conclude with a verification of Archimedes' striking claim.

First, notice that a cylinder circumscribed about a sphere of radius r itself has radius r and height $h = 2r$. The cylinder's overall surface area is the sum of the lateral surface (as in Proposition 13), as well as the circular areas of the top and bottom. Thus,

$$\begin{aligned} \text{total cylindrical surface} &= 2\pi rh + \pi r^2 + \pi r^2 \\ &= 2\pi r(2r) + 2\pi r^2 = 6\pi r^2 \\ &= \tfrac{3}{2}(4\pi r^2) \\ &= \tfrac{3}{2}(\text{spherical surface}) \end{aligned}$$

which is precisely what Archimedes meant by saying that the cylinder was "half again" the sphere in surface area.

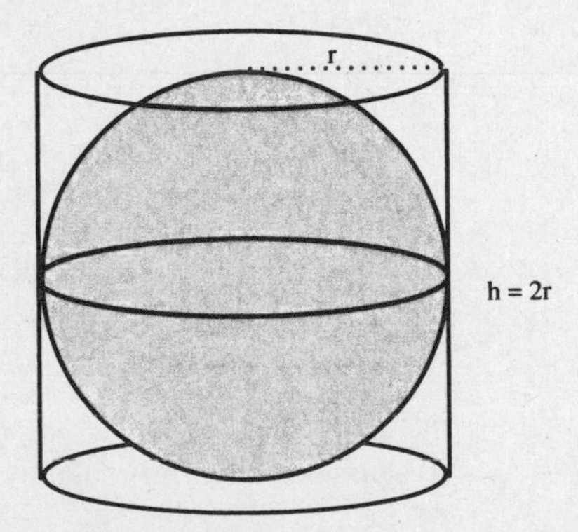

FIGURE 4.9

And what about the corresponding volumes? For a general cylinder, we have $V = \pi r^2 h$, which in this case becomes $V = \pi r^2(2r) = 2\pi r^3$. Thus,

$$\begin{aligned} \text{Cylindrical volume} &= 2\pi r^3 \\ &= \tfrac{3}{2}(\tfrac{4}{3}\pi r^3) = \tfrac{3}{2}(\text{spherical volume}) \end{aligned}$$

so that the cylinder was half again the sphere in volume.

Thus, in one concise and remarkable statement, Archimedes had linked the sphere and the cylinder. It was this link that surely accounted for the title of the treatise we are examining. That Archimedes took particular pride in this discovery was indicated by Plutarch's reference to Archimedes' choice of epitaph:

> His discoveries were numerous and admirable; but he is said to have requested his friends and relations that, when he was dead, they would place over his tomb a sphere contained in a cylinder, inscribing it with the ratio which the containing solid bears to the contained [i.e., the ratio 3:2].

Interestingly, Cicero reported in his *Tusculan Disputations* that when in Syracuse he indeed came upon Archimedes' tomb. Admittedly, "a jumble of brambles and bushes" had grown up in the area, concealing everything. But Cicero knew what he was looking for and was understandably excited when he recognized "a small column that emerged a little from the bushes: it was surmounted by a sphere in a cylinder." Having discovered the monument, he took pains to reverse the disrepair into which it had fallen. If true, Cicero had found the final resting place of the greatest of Greek mathematicians. In attempting to rescue the site from oblivion, Cicero not only paid homage to Archimedes but perhaps atoned somewhat for the brutality of his murderous Roman ancestors.

One often hears of people who are ahead of their time. By this is usually meant a man or woman who anticipates the rest of the world by a decade or perhaps even a generation. But Archimedes was doing mathematics whose brilliance would be unmatched for centuries! Not until the development of calculus in the latter years of the seventeenth century did people advance the understanding of volumes and surface areas of solids beyond its Archimedean foundation. It is certain that, regardless of what future glories await the discipline of mathematics, no one will ever again be 2000 years ahead of his or her time.

We can do no better than to end with Voltaire's fitting and quite remarkable comment on the achievements of this great mathematician: "There is more imagination in the head of Archimedes than in that of Homer."

Epilogue

One legacy of Archimedes' *Measurement of a Circle* was the quest for ever more precise estimates of the critical constant we call π. The importance of this ratio had been recognized long before Archimedes, although it was he who first subjected it to a scientific scrutiny. One interesting pre-Archimedean estimate can be inferred from a Biblical quotation about a circular "sea," that is, a large container for holding water: "Then He made the molten sea, ten cubits from brim to brim, while a line of 30 cubits measured it around" (I Kings 7:23).

From this we derive the value $\pi = C/D = 30/10 = 3.00$, an estimate which, because of its great antiquity, is quite reasonable. (Of course, here we have a bone to pick with those who regard the Bible as accurate in *all* respects, since 3.00 seriously underestimates π.)

A better ancient estimate was that of the Egyptians. In the Rhind papyrus, they used $(4/3)^4 = 256/81 = 3.1604938\ldots$ as the ratio of C to D. These and other "pre-scientific" estimates represented the first phase in the estimation of π. As we have seen, Archimedes initiated the second phase. His geometric approach, employing the perimeters of inscribed and/or circumscribed regular polygons, was the method of choice for mathematicians until the mid-seventeenth century (yet another indication that Archimedes was ahead of his time).

Around A.D. 150 the noted astronomer and mathematician Claudius Ptolemy of Alexandria provided an estimate for this ratio in his masterpiece, the *Almagest.* This extensive work was a compilation of astronomical information, from the behavior of the sun and moon, to the motions of the planets, to the nature of the fixed stars in the heavens. Obviously, the precise measurements of celestial objects required a sophisticated mathematical underpinning, and for this reason, early in the *Almagest* Ptolemy developed his Table of Chords.

He began with a circle whose diameter was divided into 120 equal parts. If each part has length p, then we can designate the diameter as $120p$, as shown in Figure 4.10. For any central angle α, Ptolemy wanted to find the length of chord AB subtended by this angle. For instance, the chord of a 60° angle is just the length of the radius, which is $60p$.

This was an easy one. Finding the chord of 42½° is far less simple. But, using some clever reasoning and showing an Archimedean knack for computation, Ptolemy generated precisely such a table for all angles from ½° up to 180° in half-degree increments.

Pertinent to our discussion, however, is the fact that he found the chord of 1° to be (in modern decimal notation) $1.0472p$. Thus, the perimeter of a regular 360-gon inscribed in this circle is 360 times as great, namely $376.992p$. Although the idea of using regular polygons is

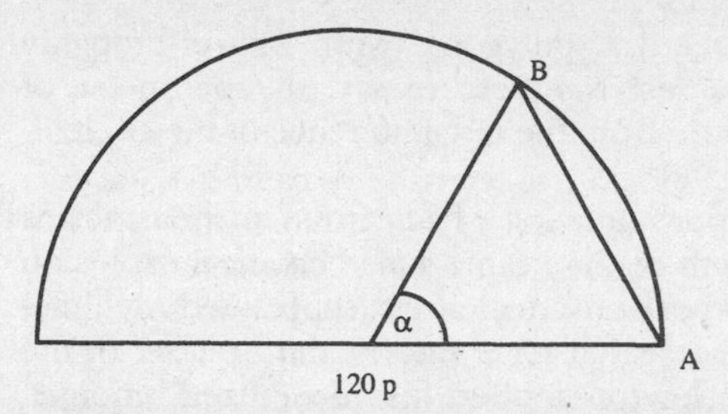

FIGURE 4.10

clearly Archimedean, Ptolemy's 360-sided figure furnished a much more accurate estimate than his predecessor's 96-gon. That is,

$$\pi = \frac{C}{D} \approx \frac{\text{perimeter of 360-gon}}{\text{diameter of circle}} = \frac{376.992p}{120p} = 3.1416$$

In the centuries that followed, advances in the calculation of π centered in the non-Western cultures of China and India, cultures with brilliant mathematical histories of their own. Thus we find the Chinese scientist Tsu Ch'ung-chih (430–501) using the estimate 355/113 = 3.14159292 . . . around A.D. 480, and the Hindu mathematician Bhāskara (1114–ca. 1185) recommending 3927/1250 = 3.1416 for accurate calculations around A.D. 1150.

When Europe finally emerged from the mathematical stagnation of the Middle Ages, the pace of discovery accelerated. By the late sixteenth century, with the work of such mathematicians as Simon Stevin (1548–1620), the modern decimal system had been established, and with it came easier, more accurate estimates of square roots. Thus, when the gifted French mathematician Francois Viète (1540–1603) tried his hand at estimating π with Archimedes' technique, he could use regular polygons of 393,216 sides to get a value accurate to nine places. This required him to follow Archimedes' lead through the 96-gon, but then to double the number of sides a dozen more times. Even Archimedes would have withered under the constraints of his number system, but the decimal notation gave Viète the opening he needed. The basic insight was still Archimedes' but Viète had better tools.

Early in the seventeenth century, a Dutch mathematician outdid all predecessors by finding π correct to 35 places. His name was Ludolph van Ceulen, and he devoted years of effort to the task. Like Viète, Ludolph combined the new decimal system with the old Archimedean strategy, although rather than starting with a hexagon and doubling its number of sides, Ludolph began with a square. By the time he was finished, he was handling regular polygons with 2^{62}—or roughly 4,610,000,000,000,000,000—sides! Needless to say, the perimeter of such a polygon differs very little from the circumference of the circle in which it is inscribed.

The classical method of approximating π had carried mathematicians far. But later in the seventeenth century came a mathematical explosion of epic proportions, one of whose advances at last supplanted Archimedes' approach and pushed the search for π into its third phase. In the late 1660s, the young Isaac Newton applied his generalized binomial theorem and newly invented method of fluxions—that is, calculus—to get a very accurate estimate of π with relative ease; this is the great theorem dealt with in Chapter 7. By 1674, Newton's rival Gottfried Wilhelm Leibniz had discovered that the series

$$1 - 1/3 + 1/5 - 1/7 + 1/9 - 1/11 + 1/13 - 1/15 + \dots$$

approaches the number $\pi/4$ as we carry the calculations ever farther along. Theoretically at least, we can extend the series of terms as far as we choose in order to get ever more accurate approximations to $\pi/4$, and consequently to π itself. It is important to note that the series we must sum here is utterly predictable in its behavior; that is, no matter where we are in the series, it is easy to determine the next term. Suddenly, then, the matter of approximating π turned from the *geometric* problem it had been with Archimedes' regular polygons to a simple ***arithmetic*** problem of adding and subtracting numerical terms. This was a major change in perspective.

Actually, the plot thickened at this point, since Leibniz's series, while it did indeed approach the number $\pi/4$, did so very slowly. For instance, even if we use the first 150 terms of the series, we get as an approximation of π only 3.1349 . . . , which is disappointingly inaccurate given the number of computations involved. It is estimated that to get 100-place accuracy with this series, one would need more than

100,000,000,000,000,000,000,000,000,000,000,000,000,000,000,000,000

terms! So, while Leibniz's series foretold the new, arithmetic approach to estimating π, it obviously had little practical use.

The promise of infinite series was soon fulfilled as mathematicians such as Abraham Sharp (1651–1742) and John Machin (1680–1751) made clever modifications that generated much more rapidly converging series. Using these adjustments, Sharp found π correct to 71 places in 1699, and Machin got 100 places seven years later. Moreover, their efforts proved far easier than those which had occupied poor Ludolph for much of his life in squeezing out 35-place accuracy. It was clear that the series approach had rendered the classical method obsolete.

Meanwhile there were developments on other fronts in mathematicians' attempts to understand this peculiar constant. Chief among these was the 1767 proof by Johann Heinrich Lambert (1728–1777) that π is an irrational number. We recall that the irrationals are those real numbers that cannot be written as the quotient of two integers—that is, the irrationals are the numbers that are not fractions. It is fairly easy to show that constants like $\sqrt{2}$ or $\sqrt{3}$ are irrational, but it took until the eighteenth century for Lambert to prove that π belonged on this list. His discovery assumes particular importance when we recall that rational numbers have decimal expansions that either terminate or exhibit a repeating pattern. For instance, the decimal for the rational number ⅛ is just .125. Alternately, the decimal for the rational ⅐ never stops, but at least it repeats in blocks of six places:

$$1/7 = .142857142857142857142857\ldots$$

If π were rational, it too would have to exhibit one of these behaviors, and thus efforts to determine its decimal expansion would, after a certain amount of time, essentially be complete. Lambert's proof that π belonged among the irrational numbers guaranteed that the computation of its decimal would forever remain unfinished business.

As if this irrationality were not already bad enough, Ferdinand Lindemann proved in 1882 that π is actually transcendental, as mentioned in Chapter 1. Not only did this discovery settle the issue of squaring the circle, but it meant that π could not emerge as any sort of elementary expression involving square roots, cube roots, and so on, of rational numbers. The results of Lambert and Lindemann showed that π is not among the "nice" numbers easily accessible to mathematical analysis. Yet the results of Archimedes from 225 B.C. had shown just as clearly that π was one of the most important numbers of all.

This history of π introduces one of the outstanding mathematicians of this century, Srinivasa Ramanujan (1887–1920). Born in India to a family of limited means, Ramanujan enjoyed none of the benefits of formal mathematical training. He was largely self-taught, and this from just a few textbooks. Ramanujan's absorption with mathematics cost him dearly in his mastery of other subjects, and his formal education ended when he was unable to pass the requisite examinations in neglected courses. By 1912, he was reduced to a clerical job in Madras, supporting himself and his wife on a mere 30 pounds per year. It would have been very easy to write him off as a failure.

Yet, despite such obstacles, this isolated genius was doing mathematical research of great originality and depth. After some urging, he wrote up a sampler of his discoveries and mailed them to three of England's foremost mathematicians. Two of them returned Ramanujan's unsolicited letter. Apparently they felt they had more pressing things to do than to respond to an unknown Indian clerk.

The third, G. H. Hardy of Cambridge University, may have been tempted to follow the same course when he opened his morning mail on January 16, 1913. Ramanujan's communication, written in poor English and containing over 100 strange formulas without proofs of any kind, seemed to be the disordered ramblings of a crackpot from halfway around the world. Hardy put the letter aside.

But, as the story goes, something about those mathematical formulas haunted him all that day. Many of the results were unlike anything Hardy had ever seen, and Hardy was among the finest mathematicians in the world. Gradually, it dawned upon him that these formulas ". . . must be true, because if they were not true, no one would have had the imagination to invent them." Indeed, when he returned to his rooms and reexamined the morning's document, Hardy realized that this was the work of an enormous mathematical talent.

Thus began the process of bringing Ramanujan to England. It was complicated by a staunch religious upbringing that placed restrictions on his mode of travel, his diet, and so on. But these problems were eventually overcome, and Ramanujan arrived at Cambridge in 1914.

There followed an extraordinary half-decade of collaboration between Ramanujan and Hardy—the latter being a sophisticated, urbane Englishman possessing the best mathematical training the world could offer; the former being a "raw talent" of incredible power who nonetheless had huge gaps in his mathematical knowledge. Sometimes Hardy had to instruct his young companion even as he would an ordinary undergraduate. At other times Ramanujan would astound him with never-before-seen mathematical results.

Among the formulas that Ramanujan devised were many that gave rapid, highly accurate approximations to π. Some of these appeared in an important 1914 paper; others were scrawled in his private notebooks (documents only now being made generally available to the world's eager mathematical community). Even the simplest of these formulas would carry us too far afield, but suffice it to say that his insights have opened lines of investigation into far more efficient estimates of π.

Unfortunately, Ramanujan's career, so improbable in its beginnings, came to a premature end. Far from home, in Cambridge during World War I, Ramanujan suffered a physical breakdown. Some attributed his decline to disease; others saw the cause as a serious vitamin deficiency brought on by his severe dietary restrictions. In the hope of recovery, he returned to India in 1919, but the familiarity of home was unable to arrest his decline. On April 26, 1920, Ramanujan died, and the world lost, at age 32, one of its mathematical legends.

We now rapidly bring our story to its modern conclusion by citing the amazing calculations of the Englishman William Shanks (1812–1882), who determined π to 707 places in 1873. Shanks had used the series approach of Machin to get this startling level of accuracy, which stood as a standard for the next 74 years. But then, in 1946, his countryman D. F. Ferguson made the startling discovery that Shanks had erred after the 527th place of his great computation. Ferguson then kindly corrected the mistake and obtained π to 710 places. For those with less of an appetite for calculations, it is difficult to imagine undertaking a *check* upon a 707-place number; more incredible is the persistence that would keep one going after finding no errors through 100 places, then 200 places, then 500 places! Yet Ferguson's inexplicable perseverance did in fact pay off.

In early 1947, the American J. W. Wrench added his own achievement to this history by publishing π to 808 places. This seemed to be a brilliant new triumph—until the indefatigable Ferguson began checking this one too. Sure enough, he found a mistake in the 723rd place of Wrench's computation. The two men then joined forces and a year later provided π correct to 808 places.

At this point, the tale enters its fourth and final phase. We have seen how people first estimated π by a sort of "rule of thumb"; next, Archimedes introduced the method of inscribed and circumscribed polygons, which prevailed until the coming of calculus when arithmetical techniques involving infinite series took over. Finally, in 1949 the computer fundamentally revolutionized the calculations. In that year, the Army's ENIAC computer found π to 2037 places. It should be stressed that this was, by modern standards, an extremely primitive machine, one which filled rooms with wires and vacuum tubes and cranked out its results with excruciating slowness. Yet even this quaint old device managed to obliterate all previous human calculations, in one leap extending the decimal estimate by two-and-a-half times beyond 22 centuries of human achievement. Not even D. F. Ferguson was going to find an error in this one. Further, as computer technology improved, the number of decimal places grew at an unbelievable pace. By 1959, there were over 16,000 places; by 1966, it had risen to a quarter of a million places, and by the late 1980s, supercomputers had pushed the expansion to somewhere over half a billion places, give or take a few million.

Yet our fragile human egos need not be too severely damaged. For, while the computers are faster at calculations than any person can hope to be, it was mathematicians who programmed the machine and thereby pointed it in the proper direction. The story of π is the story of a human, not a mechanical, triumph. And even in the late twentieth century, we must not forget that this journey had its mathematical beginnings in the short treatise *Measurement of a Circle* by the unsurpassed Archimedes of Syracuse.

Chapter 5

Heron's Formula for Triangular Area

(ca. A.D. 75)

Classical Mathematics after Archimedes

Archimedes cast a very long shadow across the mathematical landscape. Subsequent mathematicians of the classical period left their marks, but none even remotely measured up to the great Syracusan, an observation that became ever more obvious with the fall of Greek civilization and the simultaneous rise of Rome. It may be a bit simplistic, yet it is not without merit, to view Archimedes' death at the hands of a Roman centurion as a harbinger of what lay ahead. The Greeks, absorbed in their world of ideas, stood little chance before the military power of Rome; conversely, the Romans, absorbed with matters of political order and world conquest, had little regard for the abstract thinking so typical of the Greeks. Like Archimedes, the Greek tradition could not survive a new Roman order.

Some dates may be helpful. As we have seen, Syracuse fell to the Roman Marcellus in 212 B.C. The three bloody Punic Wars ended with Rome's destruction of its rival Carthage in 146 B.C., ensuring Roman control on both sides of the central Mediterranean, and that same year the last significant Greek city-state, Corinth, yielded to Roman power. A century later, Julius Caesar had conquered Gaul, and in 30 B.C., after the unsuccessful stand of Anthony and Cleopatra, Egypt fell at the hands of Octavian. Even barbarian Britain came under Roman control in A.D. 30. Rome, now officially an empire, exercised an unprecedented domination over the Western World.

With Roman conquest came their sophisticated engineering projects: bridges, roads, and aqueducts traversed the European landscape. But the abstract, pure mathematics that had so fascinated Hippocrates, Euclid, and Archimedes was not to attain its former glory.

One bright spot that remained was the great Library at Alexandria. Set in the midst of beautiful grounds and attracting the best minds from across the Mediterranean region, the Library must have been a most exciting place. It was there that a contemporary of Archimedes, the noted mathematician Eratosthenes (ca. 284–192 B.C.), spent much of his life as the chief librarian. As befits one who held such a crucial scholarly post, Eratosthenes was an enormously prolific and widely read scholar, and to him are attributed works on pure mathematics, philosophy, geography, and especially astronomy—the last including not only scholarly treatises but even a long poem, called *Hermes,* that put the fundamentals of astronomy into verse form! As with so many classical authors, most of Eratosthenes' writings are lost, and we must rely on later commentators' descriptions. But there seems to be no doubt that he was a major intellectual force in his day. Archimedes himself dedicated at least one work to Eratosthenes and regarded him as a man of considerable talent.

Among Eratosthenes' contributions was his famous "sieve," a simple technique for finding prime numbers in a straightforward, algorithmic manner. To use the sieve to strain out the primes, we begin by writing down the consecutive positive integers, starting with 2. Noting 2 as the first prime, we then cross off all subsequent multiples of 2—namely, 4, 6, 8, 10, and so on. Moving beyond 2, the next integer that has not been eliminated is 3, which must be the second prime. All of its multiples, however, can now be eliminated, so we cross off 6 (although it is already out of the running), 9, 12, 15, and so on. Then we see that 4 is already gone, so our next prime is 5; with its inclusion in the list of primes, we eliminate its multiples 10, 15, 20, 25, and so on. And so we proceed. Clearly, the numbers we cross out—being multiples of smaller integers—are not primes; thus, these composites slip through the sieve. On the other hand, primes will never fit through the mesh and thus will emerge as the only remaining numbers on our list:

2, **3**, ~~4~~, **5**, ~~6~~, **7**, ~~8~~, ~~9~~, ~~10~~, **11**, ~~12~~, **13**, ~~14~~, ~~15~~, ~~16~~, **17**, ~~18~~, **19**,
~~20~~, ~~21~~, ~~22~~, **23**, ~~24~~, ~~25~~, ~~26~~, ~~27~~, ~~28~~, **29**, ~~30~~, **31**, ~~32~~, ~~33~~, ~~34~~, ~~35~~,
~~36~~, **37**, ~~38~~, ~~39~~, ~~40~~, **41**, ~~42~~, **43**, ~~44~~, ~~45~~, ~~46~~, **47**, ~~48~~, ~~49~~, ~~50~~, ~~51~~,
~~52~~, **53**, ~~54~~, ~~55~~, ~~56~~, ~~57~~, ~~58~~, **59**, ~~60~~, **61**, ~~62~~, ~~63~~, ~~64~~, ~~65~~, ~~66~~, **67**,
~~68~~, ~~69~~, ~~70~~, **71**, ~~72~~, **73**, ~~74~~, ~~75~~, ~~76~~, ~~77~~, ~~78~~, **79**, ~~80~~, ~~81~~, ~~82~~, **83**,
~~84~~, ~~85~~, ~~86~~, ~~87~~, ~~88~~, **89**, ~~90~~, ~~91~~, ~~92~~, ~~93~~, ~~94~~, ~~95~~, ~~96~~, **97**, ~~98~~, ~~99~~

The sieve of Eratosthenes has yielded all primes below 100 in a perfectly automatic fashion. While it is clear that this process would bog down terribly in seeking all primes below, say, one hundred trillion, it should be noted that the modern computer can still get a great deal of mileage out of this ancient procedure.

Eratosthenes' best-known scientific achievement may be his reported determination of the circumference of the earth. Much has been written about this calculation, and, lacking the original treatise *On the Measurement of the Earth* in which it appeared, we are a bit uncertain as to what Eratosthenes actually did. However, tradition suggests that he used some geographic data and a very simple piece of geometry as follows:

In the Egyptian town of Syene, south of Alexandria near present-day Aswan, the sun stood directly overhead on the first day of summer. This was confirmed by the fact that an observer, peering down into a well at that moment, would be blinded by the reflected sun bouncing back up off the water. At the very same time on the very same day, a pole at Alexandria cast a small shadow. Eratosthenes observed that the angle α formed by the top of a pole and the line of its shadow was $\frac{1}{50}$ of the angle in a complete circle (see Figure 5.1). Assuming that Alexandria was due north of Syene (which was more or less correct) and that the sun was so far from the earth that its rays arrive in parallel lines (another reasonable assumption), Eratosthenes concluded by Proposition I.29 of the *Elements* that the alternate interior angle $\angle AOS$ was likewise equal to α, where O represented the center of the spherical earth, as shown in Figure 5.1. The final piece of the puzzle was the known geographical fact that the distance between the two cities had been measured as 5000 stades. Consequently we have the proportion

$$\frac{\text{distance from Syene to Alexandria}}{\text{earth's circumference}} = \frac{\text{angle } \alpha}{\text{total angle around circle}}$$

That is, 5000 stades/circumference = 1/50 and so the earth's circumference is just $50 \times 5000 = 250{,}000$ stades. At this point, the reader undoubtedly is asking, "How long is a stade?" Again, we tread on fairly

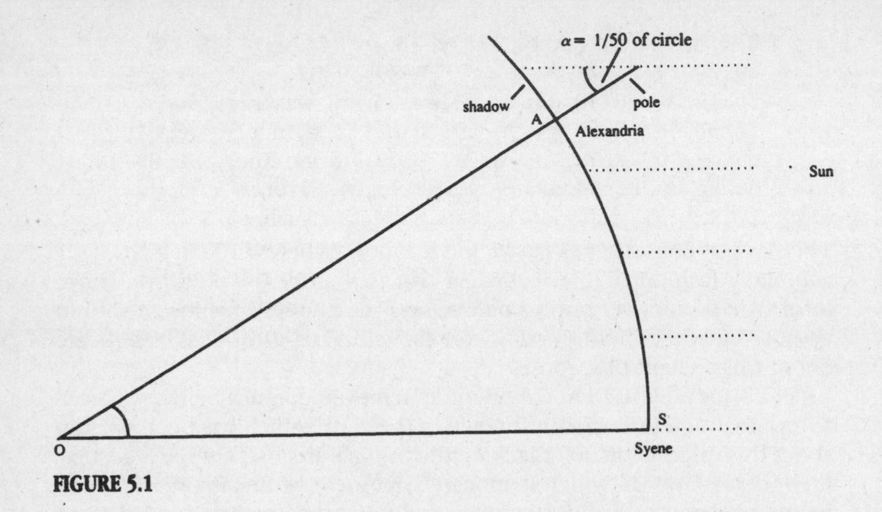

FIGURE 5.1

thin ice by citing the estimate that one stade is 516.73 feet. Using this number, we get the earth's circumference according to Eratosthenes as 129,182,500 feet, or about 24,466 miles. The figure currently accepted is 24,860 miles, so Eratosthenes was remarkably close. In fact, the estimate was so accurate that some scholars are skeptical of its authenticity, or at least agree with Sir Thomas Heath that Eratosthenes gave us "a surprisingly close approximation, however much it owes to happy accidents in the calculation."

Skepticism aside, Eratosthenes' reasoning is noteworthy not only for its cleverness but also for the striking fact that he entertained no doubts whatever that our planet was a sphere. In striking contrast, European sailors some 15 centuries later would fear plunging off the edge of a flat earth. We sometimes forget that the ancient Greeks were fully aware of the earth's spherical shape, and if later sailors kept a keen eye peeled for the horizon's edge, it was a symptom not of knowledge yet to be acquired but of knowledge lost.

Two other post-Archimedean mathematicians derserve mention. One was Apollonius (ca. 262–190 B.C.), another contemporary of Archimedes who found his way to Alexandria to work in that rich, scholarly atmosphere. There, he produced his masterpiece, the *Conics,* an extensive treatment of the so-called conic sections—the ellipse, parabola, and hyperbola (Figure 5.2). These curves had been extensively studied by Greek mathematicians, but Apollonius organized and systematized the previous work much as Euclid had done with his *Elements.* The *Conics* was written in eight books, the first four of which provided a general introduction while the remaining ones treated more specialized matters. Of these, the eighth is now entirely lost.

Even in classical times, Apollonius' work was recognized as the authoritative source on conics, and it was held in high regard when rediscovered during the Renaissance. When Johann Kepler (1571–1630) posed his groundbreaking theory that planets travel in *elliptical* orbits about the sun, the importance of the conics was affirmed. The ellipse, far from being merely a curiosity of Greek mathematics, had become the very path followed by the earth, and all of us who ride upon it. Then, almost a century later, the British scientist Edmund Halley, of comet fame, devoted years of his life to preparing the definitive edition of the *Conics*, so highly did he regard this piece of classical mathematics. Today, it stands, along with Euclid's *Elements* and the works of Archimedes, as one of the genuine landmarks of Greek mathematics.

Our final classical mathematician was responsible for the great theorem of this chapter. He was Heron of (where else?) Alexandria. In some modern books he goes by the name of "Hero," more because of the vicissitudes of translation than any pretentiousness on his part. Unfortunately, we know very little about his life, and even the century in which he lived is the subject of debate. It is certain that Heron came sometime after Apollonius, but determining more specific dates requires a talent for keen deduction most often found in detective novels. We shall go with Howard Eves and place Heron somewhere around A.D. 75.

Knowing little of his life, and not even positive that they are within 150 years of his date, scholars nonetheless have a surprising amount of information about Heron's mathematics. His interests tended to the practical rather than the theoretical, and many of his writings dealt with such useful applications as mechanics, engineering, and measurement.

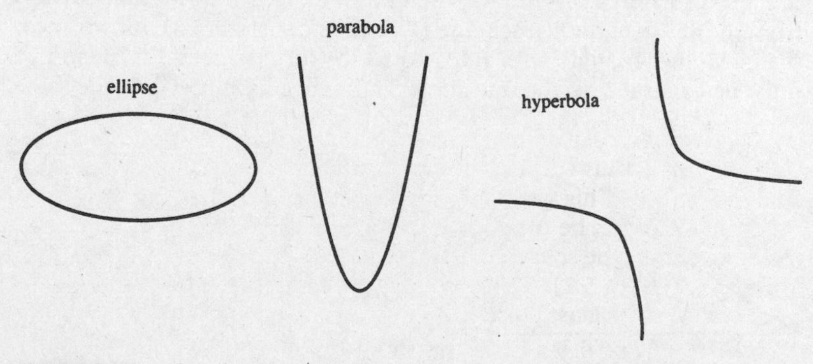

FIGURE 5.2

Such an emphasis reflects rather well the contrast between Greek and Roman interests. For instance, Heron explained in his *Dioptra* how to dig tunnels through mountains and how to measure the amount of water flowing from a spring. In another work, he answered such mundane questions as "Why does a stick break sooner when one puts one's knees against it in the middle?" or "Why do people use pincers rather than the hand to draw a tooth?"

Of interest here, however, is his proposition about the area of triangles. Like so many Heronian subjects, this clearly had its practical applicability, yet his proof was a wondrous piece of abstract geometric reasoning. It appeared as Proposition I.8 of Heron's *Metrica*, a work with quite an interesting history. Mathematicians had long known of the existence of this treatise, since it was mentioned by the commentator Eutocius in the sixth century A.D., but no traces of it existed. It seemed to be as lost as the dinosaurs when in 1894 the mathematical historian Paul Tannery happened upon a fragment of the work in a thirteenth century Parisian manuscript. Better still, two years later R. Schöne found a complete manuscript in Constantinople. By such good fortune, the *Metrica* came into modern hands.

Great Theorem: Heron's Formula for Triangular Area

Heron's formula, as noted, concerns the area of a triangle. This may seem utterly unnecessary, for the standard formula—Area = ½(base) × (height)—is simple, well known, and trivial to use. Yet it would be of little value in attacking the area of the triangle in Figure 5.3, since we have not been provided with its height.

At the outset, it is critical to note that, once a triangle's three sides are given, its area is uniquely determined. This follows immediately from the SSS congruence scheme (Euclid, Proposition I.8), for we know that any other triangle with sides equal to (for instance) 17, 25, and 26 must be congruent to the one above and hence have precisely the same

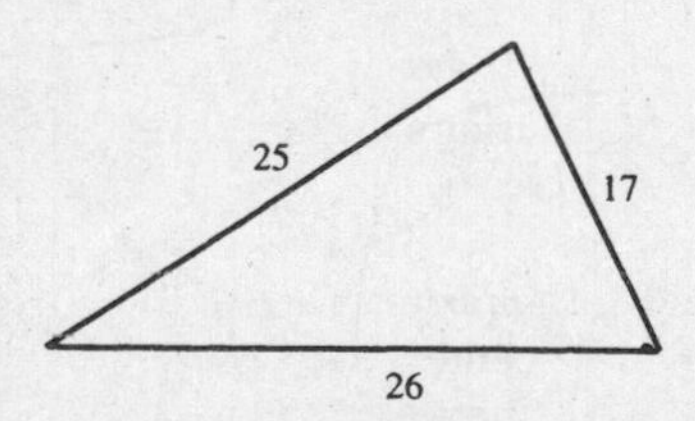

FIGURE 5.3

area. So, if we know the triangle's three sides, we also know that there is one and only one possible value for its area.

But how can we find this value? The easiest approach, today as it was two thousand years ago, is to apply Heron's formula, which in modern notation states:

> If K is the area of a triangle having sides of length a, b, and c, then
>
> $$K = \sqrt{s(s-a)(s-b)(s-c)}$$
>
> where $s = \frac{1}{2}(a+b+c)$ is the so-called "semiperimeter" of the triangle.

In Figure 5.3, $s = \frac{1}{2}(17 + 25 + 26) = 34$ and so we find

$$K = \sqrt{34(34-17)(34-25)(34-26)} = \sqrt{41616} = 204$$

Notice that, in applying Heron's formula, knowing the triangle's three sides is enough; we never need to determine its altitude.

This is a very peculiar result which, at first glance, looks like nothing if not a misprint. The presence of the square root and semiperimeter seems odd, and the formula has no intuitive appeal whatever. But it is not just its strangeness that brings it to our attention as a great theorem. Rather, it is the proof that Heron furnished, which is at once extremely circuitous, extremely surprising, and extremely ingenious. In one sense, his argument is elementary in that it uses only very simple ingredients from plane geometry—that is, only the "elements" of the subject. Yet Heron displayed an astonishing geometric virtuosity in combining these elementary pieces into a remarkably rich and elegant proof that boasts one of the best surprise endings in mathematics. As with a good Agatha Christie novel, readers of Heron's proof can be within a few lines of the end and still have no idea how the matter will be resolved. Yet we need not fear, for he ultimately brings the strands together in a wonderful climax.

Before beginning the proof itself, we need to be aware of the preliminary results upon which Heron built his argument. The first two come from Euclid.

PROPOSITION 1 The bisectors of the angles of a triangle meet at a point that is the center of the triangle's inscribed circle.

This appeared as Proposition IV.4 of Euclid's *Elements*. The point where the three angle bisectors meet—that is, the center of the triangle's inscribed circle—is called, quite appropriately, the incenter.

PROPOSITION 2 In a right-angled triangle, if a perpendicular is drawn from the right angle to the base, the triangles on each side of it are similar to the whole triangle and to one another.

Readers will recognize this as Proposition VI.8, which was examined in Chapter 3.

The next theorem, although fairly well known, appeared nowhere in Euclid. For the sake of completeness, its simple proof is included.

PROPOSITION 3 In a right triangle, the midpoint of the hypotenuse is equidistant from the three vertices.

PROOF Beginning with right triangle *BAC* (Figure 5.4), bisect side *AB* at *D* and construct *DM* perpendicular to *AB*. Drawing *MA*, we claim that $\triangle MAD$ is congruent to $\triangle MBD$, since $\overline{AD} = \overline{BD}$, $\angle ADM = \angle BDM$, and of course $\overline{DM} = \overline{DM}$. Thus the SAS congruence scheme guarantees that $\overline{MA} = \overline{MB}$ and that $\angle MAD = \angle MBD$. But we began with a right triangle. Hence,

$$\angle ACM = 1 \text{ right angle} - \angle MBD = 1 \text{ right angle} - \angle MAD = \angle MAC$$

Thus $\triangle MAC$ is isosceles, and it follows that $\overline{MC} = \overline{MA}$. Since segments *MA*, *MB*, and *MC* all have the same length, we conclude that *M*, the midpoint of the hypotenuse, is equidistant from the three vertices of our right triangle.

Q.E.D.

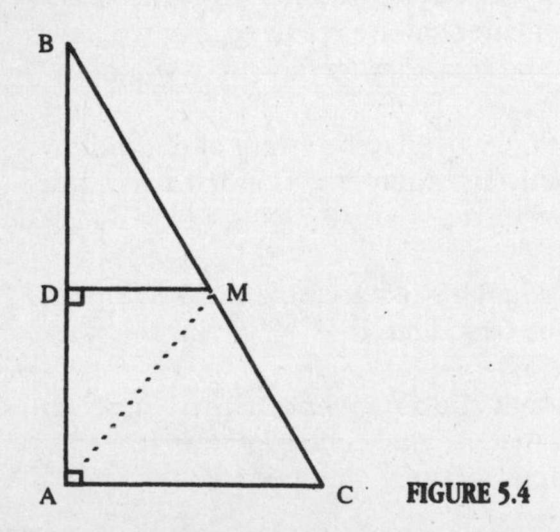

FIGURE 5.4

Our final two preliminaries deal with cyclic quadrilaterals, that is, quadrilaterals inscribed within a circle.

PROPOSITION 4 If *AHBO* is a quadrilateral with diagonals *AB* and *OH*, and if $\angle HAB$ and $\angle HOB$ are right angles (as shown in Figure 5.5), then a circle can be drawn passing through the vertices *A, O, B,* and *H.*

PROOF This fairly specialized result follows immediately from the previous one. That is, if we bisect *BH* at *M,* we observe that *M* is the midpoint of the hypotenuse of both right triangle *BAH* and right triangle *BOH.* Consequently, *M* is equidistant from points *A, O, B,* and *H,* and so a circle centered at *M* and having radius $\overline{MH}$ will pass through all four of the quadrilateral's vertices.

Q.E.D

PROPOSITION 5 The opposite angles of a cyclic quadrilateral sum to two right angles.

This appeared as Proposition III.22 of the *Elements*, and its proof was given in Chapter 3.

These five propositions may seem like a peculiar if not irrelevant tool-kit to bring to a proof about the area of general triangles. But they, along with a large dose of ingenuity, were just what Heron needed to prove the formula that now bears his name.

THEOREM For a triangle having sides of length *a, b,* and *c* and area *K,* we have $K = \sqrt{s(s-a)(s-b)(s-c)}$, where $s = \frac{1}{2}(a+b+c)$ is the triangle's semi-perimeter.

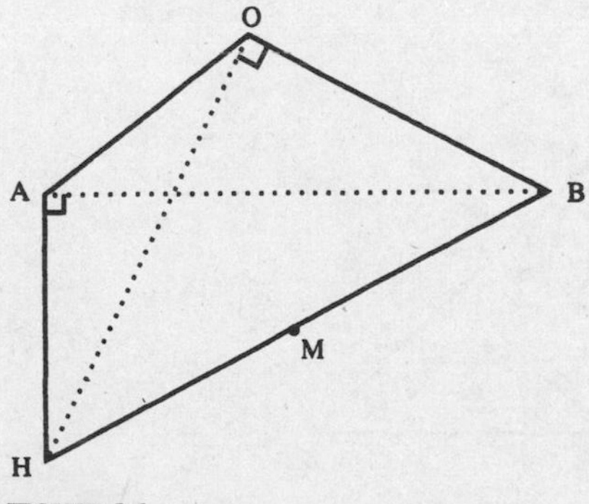

FIGURE 5.5

PROOF Let ABC be an arbitrary triangle, configured so that side AB is at least as long as the other two. To make Heron's argument flow smoothly, we shall divide it into its three main parts.

Part A Heron's very first step was something of a shocker, since he began by inscribing a circle within the triangle. This insight, to use the incenter of the triangle as a key element in determining its area, was an unexpected twist, for the properties of *circles* have no intuitive connection with the area of a rectilinear figure such as a triangle. Nonetheless, letting O be the center of the inscribed circle, and denoting its radius by r, we see that $\overline{OD} = \overline{OE} = \overline{OF} = r$, as shown in Figure 5.6.

Now we apply the simple formula for triangular area to get:

$$\text{Area }(\triangle AOB) = \tfrac{1}{2}(\text{base}) \times (\text{height}) = \tfrac{1}{2}(\overline{AB}) \times (\overline{OD}) = \tfrac{1}{2}cr$$
$$\text{Area }(\triangle BOC) = \tfrac{1}{2}(\text{base}) \times (\text{height}) = \tfrac{1}{2}(\overline{BC}) \times (\overline{OE}) = \tfrac{1}{2}ar$$
$$\text{Area }(\triangle COA) = \tfrac{1}{2}(\text{base}) \times (\text{height}) = \tfrac{1}{2}(\overline{AC}) \times (\overline{OF}) = \tfrac{1}{2}br$$

Thus, $K = \text{Area }(\triangle ABC) = \text{Area }(\triangle AOB) + \text{Area }(\triangle BOC) + \text{Area }(\triangle COA)$, or

$$K = \tfrac{1}{2}cr + \tfrac{1}{2}ar + \tfrac{1}{2}br = r\left(\frac{a+b+c}{2}\right) = rs$$

Here we see Heron's link between the triangle's area, K, and its semiperimeter, s. While this suggests that we are on the right track, much more work awaits.

Part B We again refer to Figure 5.6 and recall from our first preliminary that the process of inscribing circles began by bisecting the triangle's

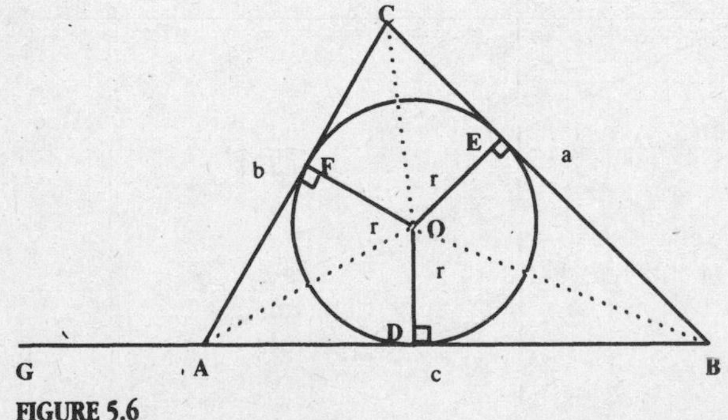

FIGURE 5.6

three angles. Thus, $\triangle ABC$ has been decomposed into three pairs of congruent triangles, namely

$$\triangle AOD \cong \triangle AOF, \quad \triangle BOD \cong \triangle BOE, \quad \text{and} \quad \triangle COE \cong \triangle COF,$$

where in each case the congruence follows by AAS (Euclid, Proposition I.26). Then, by corresponding parts, we have

$$\overline{AD} = \overline{AF}, \quad \overline{BD} = \overline{BE}, \quad \text{and} \quad \overline{CE} = \overline{CF}$$

$$\text{while} \quad \angle AOD = \angle AOF, \quad \angle BOD = \angle BOE, \quad \text{and} \quad \angle COE = \angle COF$$

At this point, Heron extended the triangle's base AB to the point G, where $\overline{AG} = \overline{CE}$. He then argued that

$$\begin{aligned}
\overline{BG} &= \overline{BD} + \overline{AD} + \overline{AG} = \overline{BD} + \overline{AD} + \overline{CE} \quad \text{by construction} \\
&= \tfrac{1}{2}(2\overline{BD} + 2\overline{AD} + 2\overline{CE}) \\
&= \tfrac{1}{2}[(\overline{BD} + \overline{BE}) + (\overline{AD} + \overline{AF}) + (\overline{CE} + \overline{CF})] \quad \text{by congruence} \\
&= \tfrac{1}{2}[(\overline{BD} + \overline{AD}) + (\overline{BE} + \overline{CE}) + (\overline{AF} + \overline{CF})] \\
&= \tfrac{1}{2}[\overline{AB} + \overline{BC} + \overline{AC}] = \tfrac{1}{2}(c + a + b) = s
\end{aligned}$$

Consequently, Heron's segment BG had as its length the triangle's semiperimeter, albeit "straightened out." Apparently, Heron wanted to have the semiperimeter before him, all in one piece.

Knowing that $\overline{BG} = s$, we easily derive

$$s - c = \overline{BG} - \overline{AB} = \overline{AG}$$

$$\begin{aligned}
s - b &= \overline{BG} - \overline{AC} \\
&= (\overline{BD} + \overline{AD} + \overline{AG}) - (\overline{AF} + \overline{CF}) \\
&= (\overline{BD} + \overline{AD} + \overline{CE}) - (\overline{AD} + \overline{CE}) = \overline{BD}
\end{aligned}$$

since $\overline{AD} = \overline{AF}$ and $\overline{AG} = \overline{CE} = \overline{CF}$. Likewise,

$$\begin{aligned}
s - a &= \overline{BG} - \overline{BC} \\
&= (\overline{BD} + \overline{AD} + \overline{AG}) - (\overline{BE} + \overline{CE}) \\
&= (\overline{BD} + \overline{AD} + \overline{CE}) - (\overline{BD} + \overline{CE}) = \overline{AD}
\end{aligned}$$

since $\overline{BD} = \overline{BE}$ and $\overline{AG} = \overline{CE}$.

In short, the semiperimeter s and the quantities $s - a$, $s - b$, and $s - c$ all appear as particular segments in the diagram. Again, this is suggestive since these are the components of the formula we seek to prove. What remained for Heron was to assemble these components to complete his argument.

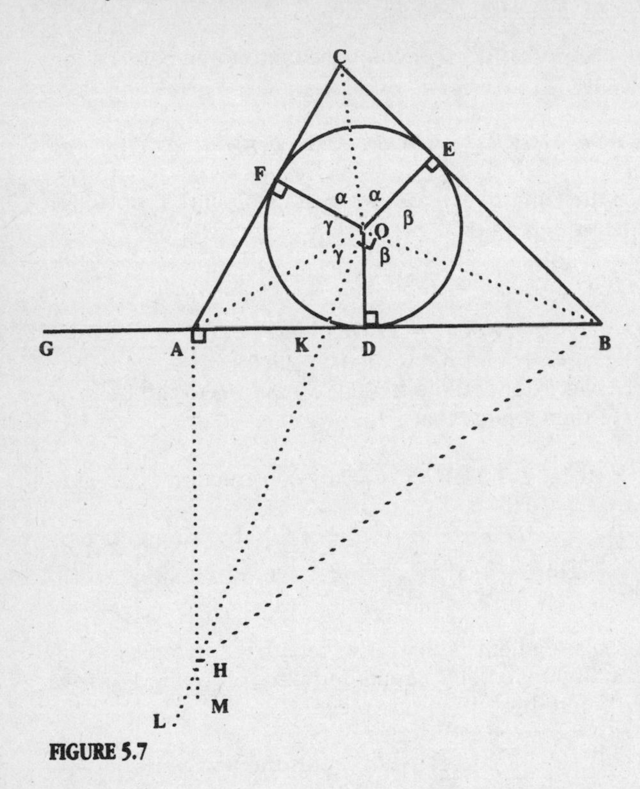

FIGURE 5.7

Part C Again we begin with $\triangle ABC$ and its inscribed circle, but we now need an extended diagram to illustrate Heron's reasoning (Figure 5.7). He drew *OL* perpendicular to *OB*, cutting *AB* at *K*. Next, he constructed *AM* perpendicular to *AB* meeting *OL* at the point *H*, and finally he drew *BH*.

The resulting quadrilateral *AHBO* should look familiar. By Proposition 4, it is, in fact, a cyclic quadrilateral and so, by Proposition 5, we know that its opposite angles sum to two right angles. That is,

$$\angle AHB + \angle AOB = \text{two right angles}$$

Now examine the angles about the incenter *O*. By the congruences from part *B*, these reduce to three pairs of equal angles, so that

$$2\alpha + 2\beta + 2\gamma = \text{four right angles} \qquad \text{or equivalently}$$
$$\alpha + \beta + \gamma = \text{two right angles}$$

But $\beta + \gamma = \angle AOB$, and thus $\alpha + \angle AOB =$ two right angles $= \angle AHB + \angle AOB$. Hence $\alpha = \angle AHB$, a seemingly insignificant fact that will turn out to be crucial in what follows.

Heron next observed that $\triangle COF$ is *similar* to $\triangle BHA$, for $\angle CFO$ and $\angle BAH$ are both right angles and, by the previous comment, $\alpha = \angle AHB$. From this similarity, we derive the proportion

$$\frac{\overline{AB}}{\overline{AH}} = \frac{\overline{CF}}{\overline{OF}} = \frac{\overline{AG}}{r}$$

since $\overline{CF} = \overline{AG}$ and $\overline{OF} = r$. This is equivalent to the following equation, which we shall call (*).

$$\frac{\overline{AB}}{\overline{AG}} = \frac{\overline{AH}}{r} \qquad (*)$$

Heron noted that $\triangle KAH$ is likewise similar to $\triangle KDO$, for $\angle KAH$ and $\angle KDO$ are both right, while vertical angles $\angle AKH$ and $\angle DKO$ are equal. This similarity yields:

$$\frac{\overline{AH}}{\overline{AK}} = \frac{\overline{OD}}{\overline{KD}} = \frac{r}{\overline{KD}} \qquad \text{and thus} \qquad \frac{\overline{AH}}{r} = \frac{\overline{AK}}{\overline{KD}}$$

Combining this last line with equation (*) yields the key result, which we shall call (**).

$$\frac{\overline{AB}}{\overline{AG}} = \frac{\overline{AK}}{\overline{KD}} \qquad (**)$$

At this point, Heron's readers may be forgiven for suspecting that the mathematician was adrift, wandering aimlessly through an unending series of similar triangles. This feeling is by no means dispelled by his next step, which examined yet another pair of similar triangles.

Heron looked at $\triangle BOK$ with altitude $\overline{OD} = r$. By preliminary Proposition 2, we know that $\triangle KDO$ is similar to $\triangle ODB$ and thus

$$\frac{\overline{KD}}{r} = \frac{r}{\overline{BD}} \quad \text{or simply} \quad (\overline{KD})(\overline{BD}) = r^2 \qquad (***)$$

(The Greeks would simply say that r is the "mean proportional" between magnitudes $\overline{KD}$ and $\overline{BD}$.)

At this point, Heron added 1 to each side of equation (**) to get

$$\frac{\overline{AB}}{\overline{AG}} + 1 = \frac{\overline{AK}}{\overline{KD}} + 1$$

Over common denominators this becomes

$$\frac{\overline{AB} + \overline{AG}}{\overline{AG}} = \frac{\overline{AK} + \overline{KD}}{\overline{KD}} \quad \text{or simply} \quad \frac{\overline{BG}}{\overline{AG}} = \frac{\overline{AD}}{\overline{KD}}$$

In this last equation, multiplication of the left-hand fraction by $\overline{BG}/\overline{BG}$ and the right-hand one by $\overline{BD}/\overline{BD}$ will certainly maintain the equality and yield

$$\frac{(\overline{BG})(\overline{BG})}{(\overline{AG})(\overline{BG})} = \frac{(\overline{AD})(\overline{BD})}{(\overline{KD})(\overline{BD})} \quad \text{and so}$$

$$\frac{(\overline{BG})^2}{(\overline{AG})(\overline{BG})} = \frac{(\overline{AD})(\overline{BD})}{r^2} \quad \text{by (***)}$$

Cross-multiplying the previous equation gives us

$$r^2 (\overline{BG})^2 = (\overline{AG})(\overline{BG})(\overline{AD})(\overline{BD})$$

But at last Heron was ready to assemble this multitude of pieces to come rapidly and spectacularly to his desired end. We need only recognize that the components of this last equation above are precisely the segments identified in Part B. Making the substitution gives:

$$r^2s^2 = (s-c)(s)(s-a)(s-b) = s(s-a)(s-b)(s-c)$$

$$\text{and so} \qquad rs = \sqrt{s(s-a)(s-b)(s-c)}$$

But we recall from Part A that if K is the area of our triangle, then $rs = K$. Thus a final substitution gives Heron's formula:

$$K = \sqrt{s(s-a)(s-b)(s-c)}$$

Q.E.D.

Thus ended one of the cleverest proofs from elementary geometry, whose unexpected and apparently random wanderings were in fact always directed toward the desired end. This is certainly the most convoluted proof we have encountered to date. It is difficult to imagine the mental gyrations that would have led Heron to devise such a spectacularly devious argument. Recalling that he is sometimes known by the shorter version of his name, we may perhaps label his performance Hero-ic indeed.

Epilogue

Historians have turned up a curious fact about this remarkable formula. In an old Arabic manuscript written centuries after Heron, the Islamic scholar Abu'l Raihan Muh. al-Biruni credited this result not to Heron but to the illustrious Archimedes himself. We have no Archimedean writings to support this claim, but so extraordinary was his intellect that such a theorem would certainly have been within his reach.

On the other hand, for reasons of sentiment more than historical accuracy, it may be best to allow Heron his moment in the sun. Crediting this result to Archimedes rather than Heron seems unnecessarily generous to the former, whose reputation is already unsurpassed among classical mathematicians, and seems unnecessarily cruel to the latter, whose reputation rests so much upon it.

As noted, Heron's formula has any number of practical applications. Surveyors who know the lengths of a three-sided lot can easily compute the area, and lots with four or more sides can easily be decomposed into triangular fragments for area determinations. But Heron's formula can also be used to yield an old friend, as we shall now see.

Suppose we have a right triangle with hypotenuse of length a and legs of length b and c, as shown in Figure 5.8. Here the semiperimeter is

$$s = \frac{a + b + c}{2} \qquad \text{and we find}$$

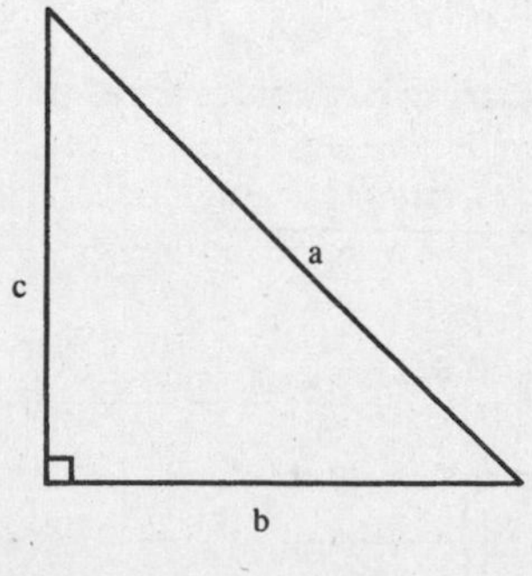

c a b

FIGURE 5.8

$$s - a = \frac{a + b + c}{2} - a = \frac{a + b + c}{2} - \frac{2a}{2} = \frac{-a + b + c}{2}$$

Similarly

$$s - b = \frac{a - b + c}{2} \quad \text{and} \quad s - c = \frac{a + b - c}{2}$$

Further, a bit of algebra confirms that

$$\begin{aligned}(a + b + c)(-a + b + c)&(a - b + c)(a + b - c)\\ &= [(b + c) + a][(b + c) - a][a - (b - c)][a + (b - c)]\\ &= [(b + c)^2 - a^2][a^2 - (b - c)^2]\\ &= a^2(b + c)^2 - (b + c)^2(b - c)^2 - a^4 + a^2(b - c)^2\end{aligned}$$

which simplifies to $2a^2b^2 + 2a^2c^2 + 2b^2c^2 - (a^4 + b^4 + c^4)$.

Thus, when we return to Heron's formula, we get the area of the triangle to be

$$\begin{aligned}K &= \sqrt{s(s - a)(s - b)(s - c)}\\ &= \sqrt{\left(\frac{a + b + c}{2}\right)\left(\frac{-a + b + c}{2}\right)\left(\frac{a - b + c}{2}\right)\left(\frac{a + b - c}{2}\right)}\\ &= \sqrt{\frac{2a^2b^2 + 2a^2c^2 + 2b^2c^2 - (a^4 + b^4 + c^4)}{16}}\end{aligned}$$

On the other hand, the area of the triangle above can be easily determined as

$$K = \tfrac{1}{2}(\text{base}) \times (\text{height}) = \tfrac{1}{2}bc$$

Equating these two expressions for K and squaring both sides gives us

$$\frac{b^2c^2}{4} = \frac{2a^2b^2 + 2a^2c^2 + 2b^2c^2 - (a^4 + b^4 + c^4)}{16}$$

which, with a cross-multiplication, becomes

$$4b^2c^2 = 2a^2b^2 + 2a^2c^2 + 2b^2c^2 - (a^4 + b^4 + c^4)$$

Now, taking all terms to the left side and deftly grouping them yields

$$(b^4 + 2b^2c^2 + c^4) - 2a^2b^2 - 2a^2c^2 + a^4 = 0 \qquad \text{or simply}$$

$$(b^2 + c^2)^2 - 2a^2(b^2 + c^2) + a^4 = 0 \qquad \text{or even more simply}$$

$$[(b^2 + c^2) - a^2]^2 = 0$$

This, at long last, allows us to conclude that $(b^2 + c^2) - a^2 = 0$, which reduces to the familiar-looking $a^2 = b^2 + c^2$. So, Heron's formula provides us with another proof of the Pythagorean theorem. Of course, this proof is incredibly more complicated than is necessary—rather like traveling from Boston to New York by way of Spokane—but nonetheless it is remarkable to find the Pythagorean theorem emerging, albeit rather indirectly, from Heron's curious result.

Euclid, Archimedes, Eratosthenes, Apollonius, Heron—these and many other mathematicians had associations with the School of Alexandria, a center of scientific work that prevailed century after century during classical times. But even as the Roman Empire was not immortal, neither was this great facility.

The Alexandrian Library remained active from its founding around 300 B.C. until its closing by Christians in A.D. 529 (they objected to its huge collection of pagan documents) and its ultimate burning by the Arabs in A.D. 641. While many items were saved from the flames, much of classical civilization was forever lost in this conflagration. As with other lost monuments of the past—the treasures from the Great Pyramid of Cheops or the Temple of David in Jerusalem or the nearby Pharos of Alexandria—archaeologists today can only shrug their shoulders in numb frustration at the knowledge and beauty that are irretrievably gone.

The focus of mathematical activity, so long centered at Alexandria, had shifted. From A.D. 641 and for many, many centuries afterward, Arabian mathematicians would be the guardians of classical scholarship as well as mathematical innovators in their own right. Of course, the story of the Islamic Empire must begin with the life of Mohammed (A.D. 570–632), who rose from obscurity to become one of the pivotal figures in world history. A century and a half after Mohammed's death in Jerusalem, the religion he had founded stretched from India, through Persia and the Middle East, across northern Africa, and on into southern Spain. As they spread geographically, the Islamic scholars eagerly assimilated the knowledge of the many civilizations with which they came in contact.

Among this knowledge was the mathematics of the Hindus in India, from which the so-called "Hindu-Arabic" numeral system arose. This system was so superior to that of the Romans that it has relegated the latter to clock faces, copyright dates, and Super Bowls. Had they done nothing else, the Arabs would long be remembered for promulgating this most useful numeral system.

But of course, they did much more. In the early 800s, the Arabs began translating the Greek classics, as well as providing helpful commentaries on these works. The *Elements* was translated in A.D. 800, and Ptolemy's classic *Syntaxis Mathematica* followed a few decades later. This latter work, from around A.D. 150, was the ultimate astronomical treatise of the classical world. Mimicking Euclid, it was composed of 13 books, including those on eclipses, on the sun, on the planets, and on the stars, as well as a Table of Chords mentioned in the Epilogue to Chapter 4. Ptolemy also spelled out in great detail his model of the solar system, an earth-centered model that would serve the needs both of science and of the human ego for 1400 years until the coming of a Polish thinker named Copernicus. The Arabs held Ptolemy's work in such high regard that they called it "Al magiste"—Arabic for "The Greatest"—and thus it is that we know it today under the title of the *Almagest*.

Somewhat later, the great scholar Tâbit ibn Qorra (826–901) succeeded in producing fine translations of Archimedes and Apollonius and rendering a very faithful translation of the *Elements*. The center of such Arabian scholarship was the city of Baghdad, in present-day Iraq, where there was established "The House of Wisdom," a beehive of scholarly activity that counted among its members a host of astronomers, mathematicians, and translators. The center of the mathematical world—having previously resided at Plato's Academy and the Library of Alexandria—had now shifted to Baghdad, where it would remain for a very long time.

Among the most important of the Arabian mathematicians was Mohammed ibn Mûsâ al-Khowârizmî (ca. A.D. 825). Borrowing from both East and West—that is, from both the Hindu mathematician Brahmagupta and the Greeks we have encountered thus far—al-Khowârizmî produced a treatise on algebra and arithmetic that would prove very influential. In it, al-Khowârizmî illustrated the solution not only of linear (first-degree) equations but also of quadratic (second-degree) ones. That is, for the quadratic equation $ax^2 + bx + c = 0$, the solutions are

$$x = \frac{-b \pm \sqrt{b^2 - 4ac}}{2a}$$

Al-Khowârizmî's presentation of this result was entirely verbal, without the concise algebraic symbolism we now employ. But if he did not give algebra its symbols, he gave it its name, at least indirectly. His major treatise was titled *Hisâb al-Jabr w'al muqâbalah.* When this was translated into Latin four centuries later, the title emerged as *Ludus algebrae et almucgrabalaeque,* and the shortened term "algebra" eventually stuck.

There is debate as to the ultimate contributions of Arabian mathematicians. On the one hand, while they studied the work of such giants as Euclid and Archimedes, they never duplicated their glories. Nowhere in the Islamic works do we find the kind of quantum leaps of mathematical knowledge that so characterized the succession of Greek scholars. In particular, the Arabs simply did not regard "proof" as being at the heart of their mathematics, and in this sense mimicked the pre-Greek civilizations of the Near East. Because the Islamic mathematicians put less emphasis on proving their results in complete generality, no Arabian great theorem appears here.

On the other hand, the Arab mathematicians did popularize a highly useful number system and contributed significantly to the problem of solving equations of various degrees. Moreover, in the words of Howard Eves, they were the "custodians of much of the world's intellectual possessions" for the centuries when Europe slept. Without this great service, much of our knowledge of classical culture generally, and classical mathematics in particular, might have been forever lost.

Eventually, the Arabs would relinquish their custodianship of Euclid and Archimedes, and these works would filter back into Europe. A chief impetus was, of course, the series of Crusades from the late-eleventh to mid-thirteenth centuries, in which the relatively backward Christian West met the relatively more sophisticated Islamic East. The Europeans failed to wrest the Holy Land from Moslem dominion but did return open-eyed at the high level of learning that existed among their enemies.

Perhaps more significant was the Christian conquest of the Moors in Spain and Sicily. The great Spanish city of Toledo fell to the Christians in 1085, and Sicily was conquered a few years later. When the Europeans entered these defeated territories, they found the books and documents of the vanquished Arabs. With an unimagined world of knowledge at their fingertips and the ability to study it at their leisure, the Europeans began to discover the scholarship not only of their Islamic adversaries but of their classical ancestors. The effect was dramatic.

Much of the impact of these classics—works by Plato and Aristotle and of course Euclid—was felt in the emerging universities of Italy. The first was founded at Bologna in 1088 and others soon followed at Padua, Naples, Milan, and elsewhere. Over the next century or two, the intellectual climate in Italy rose from its Medieval depths toward the heights we now call the Renaissance.

And it was in sixteenth-century Italy that the Arabic transmission of classical culture combined with the awakening talent of the Italian scholars to produce our next great theorem: the bizarre and incredible tale of the solution of the cubic equation by Gerolamo Cardano of Milan.

Chapter 6

Cardano and the Solution of the Cubic (1545)

A Horatio Algebra Story

Without question, the last decades of the fifteenth century marked a time of great intellectual excitement in Europe. Western civilization had clearly awakened from the slumber of the Middle Ages. Johannes Gutenberg had invented his marvelous printing press in 1450, and books became available as never before. Universities at Bologna, Paris, Oxford, and elsewhere had become legitimate centers of higher education and scholarship. In Italy, Raphael and Michaelangelo were beginning extraordinary artistic careers while their older countryman, Leonardo da Vinci, was giving meaning to the term Renaissance man.

It was not just the intellectual world whose horizons were expanding. In the year 1492, Christopher Columbus, a Genoa native, had discovered a new world far across the Atlantic Ocean. As much as anything, this discovery of the Americas stood as proof that contemporary civilization could extend the frontiers of knowledge beyond even the glorious legacy of the classical world. As the fifteenth century waned, there could be no doubt that Europe was on the threshold of great things.

And so it was in mathematics. In the year 1494, the Italian Luca Pacioli (ca. 1445–1509) produced a volume titled *Summa de Arithmetica.* In it, Pacioli treated the standard mathematics of his day, with emphasis on solving both linear and quadratic equations. Interestingly, he flirted with a primitive symbolic algebra by using *co* to denote the unknown quantity in his equations. This was short for *cosa,* the Italian word for "thing"—that is, the thing to be determined. It would be a century or more before algebra evolved into the symbolic system that we recognize today, but *Summa de Arithmetica* had taken a step in this direction.

Pacioli's assessment of the cubic equation—that is, an equation of the form $ax^3 + bx^2 + cx + d = 0$—was decidedly pessimistic. He had no idea how to solve the general cubic and expressed the belief that such a solution was as impossible, given the state of mathematics, as squaring the circle. This observation, actually something of a challenge laid before the Italian mathematical community, set the stage for the remarkable tale that surrounds our next great theorem: the sixteenth-century Italian algebraists and their quest for the solution of the cubic.

The story begins with Scipione del Ferro (1465–1526) of the University of Bologna. Taking up Pacioli's challenge, the talented del Ferro discovered a formula that solved the so-called "depressed cubic." This is a third-degree equation that lacks its second degree, or quadratic, term. That is, the depressed cubic looks like $ax^3 + cx + d = 0$. Usually, we prefer to divide through by a and move the constant term to the right-hand side of the equal sign, so as to convert the depressed cubic to its standard form

$$x^3 + mx = n$$

Renaissance Italians called this "cube and cosa equals number," for obvious reasons. Although he had mastered only this particular kind of cubic, del Ferro's algebraic advance was significant, and we would expect him to have spread the word of his triumph far and wide. Actually, he did nothing of the sort. The cubic's solution he kept an absolute secret!

To understand such behavior—almost incomprehensible in the "publish or perish" world of today—we must consider the nature of the Renaissance university. There, academic appointments were by no means secure. Along with patronage and political influence, continued service depended on the ability to prevail in public challenges that could be issued from any quarter at any time. Mathematicians like del Ferro always had to be ready to do scholarly battle with challengers, and the consequence of a public humiliation could be disastrous to one's career.

Thus, a major new discovery was a powerful weapon. Should an opponent appear with a list of problems to be solved, del Ferro could counter with a list of depressed cubics. Even if del Ferro were stumped by some of his challenger's problems, he could feel confident that his cubics, baffling to all but himself, would guarantee the downfall of his unfortunate adversary.

Scipione apparently did a good job of keeping his solution secret throughout his life, and it was only on his deathbed that he passed it along to his student Antonio Fior (ca. 1506–?). Although Fior was not so good a mathematician as his mentor, he rashly went on the offensive with his new-found weapon and in 1535 leveled a challenge at the noted Brescian scholar Niccolo Fontana (1499–1557).

An unfortunate childhood calamity had shaped Fontana's life. During the French attack on his home town in 1512, a soldier, sword in hand, had delivered a savage, slashing wound to the face of young Niccolo. According to legend, the boy survived only because a dog licked the horrible gash. But if the medicinal effects of canine saliva saved his life, they could not save his speech. So disfigured was Niccolo Fontana that he could no longer speak with clarity. Tartaglia—the Stammerer—became his nickname, and it is by this rather cruel epithet that he is best known today.

Physical deformities aside, Tartaglia was a gifted mathematician. In fact, he boasted that he could solve cubics of the form $x^3 + mx^2 = n$—that is, cubics missing their linear terms—although Fior doubted that Tartaglia had such a method. When the challenge from Fior arrived, Tartaglia sent him a list of 30 problems covering various mathematical topics. By contrast, Fior had provided a list of 30 "depressed cubics" and thereby placed Tartaglia in a bind. It was clearly a case of Fior's putting all his eggs into one basket; Tartaglia was either going to get a score of 0 or of 30 depending on whether or not he found the secret.

Not surprisingly, Tartaglia began a frantic, round-the-clock attack on the depressed cubic. His frustrations mounted as the days passed and the critical deadline approached. Then, on the night of February 13, 1535, with time almost exhausted, Tartaglia discovered the solution. His intense efforts had paid off. He now could solve all of Fior's problems with ease, while his less gifted challenger turned in a dismal performance of his own. In a great public triumph, Tartaglia prevailed brilliantly. His reward was to have been 30 lavish banquets provided by the hapless Fior, but Tartaglia, in a gesture of magnanimity, relieved his opponent of this commitment. The monetary savings to Fior must have been of little value as compared to the total disgrace he had suffered; he quietly faded from the picture.

But then entered perhaps the most bizarre character in the whole history of mathematics, Gerolamo Cardano (1501–1576) of Milan. Cardano had heard of the challenge and desired to learn more of the wonderful techniques of Tartaglia, the master of the cubic equation. Rather boldly, Cardano asked the Brescian to divulge the secret, and from there the story took unexpected and remarkable turns.

Before following it to its conclusion, however, we should pause to examine the extraordinary life of Gerolamo Cardano. We are fortunate to have a first-person account in his autobiography *De Vita Propria Liber (The Book of My Life)* written in 1575. This book is awash with Cardano's recollections, peeves, and superstitions, not to mention a wealth of extremely peculiar anecdotes. More than most autobiographies, this one must be regarded skeptically; even so, it gives us a revealing glimpse of his turbulent life.

Cardano began with a brief discussion of his forebears. His family tree may have included Pope Celestino IV, not to mention a distant cousin Angiolo, who, at the venerable age of eighty

> begot sons—infants feeble as if with their father's senility . . . The eldest of these sons has lived to be seventy, and I hear that some of his children became giants.

Then, in a chapter called "My Nativity," Cardano revealed that "although various abortive medicines—as I have heard—were tried in vain" he survived, only to be "literally torn from my mother's womb." This experience left him nearly dead, and a bath of warm wine was required to bring the infant Gerolamo back to life. It appears that Cardano may have been illegitimate, thus explaining his unwelcome arrival, and the associated stigma played a key role in his life's story.

With such a shaky start, it should come as no surprise that Cardano was plagued with infirmities throughout his life. In his autobiography, he never hesitated to describe these afflictions, often in complete if not disgusting detail. He told of violent heart palpitations, of fluids oozing from the stomach and chest, of ruptures and hemorrhoids, not to mention a disease characterized by "an extraordinary discharge of urine" yielding up to 100 ounces (nearly a gallon) per day. He recorded an intense fear of high places, as well as "of places where there is any report of mad dogs having been seen." He experienced years of sexual impotence, which lasted until just before his marriage (certainly an example of good timing). It was not unusual for Cardano to experience eight consecutive nights of insomnia; at such times there was little he could do but "get up, walk around the bed, and count to a thousand many times."

On those rare occasions when he was not suffering from one of his horrible ailments, Cardano would consciously inflict pain upon himself. He did so because "I considered that pleasure consisted in relief following severe pain" and, when not suffering physically, "a certain mental anguish overcomes me, so grievous that nothing could be more distressing." Consequently,

> I have hit upon a plan of biting my lips, of twisting my fingers, of pinching the skin of the tender muscles of my left arm until the tears come.

Cardano was saying, more or less, that these self-inflicted tortures were desirable because it felt so good when he stopped.

Fragile physical (and mental) health was not his only problem. After compiling an excellent record at the University of Padua on the way to becoming a physician, Cardano was refused permission to practice medicine in his home of Milan. This refusal may have been due to his reputed illegitimacy or to his grating and bizarre personality, but whatever its cause, it marked one of the low points in a life notable for its ups and downs.

Rejected by Milan, Cardano moved to the small town of Sacco, near Padua, where he practiced medicine in the bucolic, if somewhat limiting, confines of country life. One night in Sacco, he dreamt of a beautiful woman in white. As one who put great stock in the meaning of dreams, he was thus strongly affected when, some time later, he encountered a woman exactly matching his dream apparition. At first, the poor Cardano despaired at the impossibility of courting her:

> If I, a pauper, marry a wife who has no dot save a troop of dependent brothers and sisters, I'm done for! I can scarcely pay my expenses as it is! If I should attempt an abduction, or try to seduce her, there would be plenty to spy upon me.

Still, his love made marriage irresistible. In 1531, he married Lucia Bandarini, the woman of his dreams.

As this episode suggests, dreams, omens, and portents figured prominently throughout Cardano's life. He was an ardent astrologer, a wearer of amulets, and a seer of visions who predicted the future from thunderstorms. In addition, he often felt the presence of a protective spirit, or guardian angel, as he remarked in his autobiography:

> Attendant or guardian spirits . . . are recorded as having favored certain men constantly—Socrates, Plotinus, Synesius, Dio, Flavius Josephus—and I include myself. All, to be sure, lived happily save Socrates and me . . .

Apparently, he did not hesitate to carry on lively conversations with his attendant spirit. Says Oystein Ore, Cardano's twentieth century biographer, "In the face of such tales it is no wonder that some of his contemporaries believed that he was not in his right mind."

Another of his life-long interests was gambling. Cardano regularly indulged in games of chance, often earning substantial sums to supplement his income. Contritely, he acknowledged in his autobiography

> . . . as I was inordinately addicted to the chess-board and the dicing table, I know that I must rather be considered deserving of the severest censure. I gambled at both for many years; and not only every year, but—I say it with shame—every day.

Fortunately, Cardano subjected this vice to a scientific scrutiny. His resulting *Book on Games of Chance*, published posthumously in 1663, was the first serious treatise on the mathematics of probability.

And so, casting horoscopes, constantly gambling, beginning a family, Gerolamo Cardano spent the years from 1526 to 1532 in Sacco. But neither his pocketbook nor his ego could endure the small-town atmosphere for long, and by 1532 Cardano, with wife Lucia and son Giambattista, was back in Milan, still forbidden to practice medicine and ultimately consigned to the poorhouse.

Then, at last, fortune smiled upon him. Cardano began giving lectures on popular science that were especially well received by the educated and nobility. He wrote successful treatises on topics ranging from medicine to religion to mathematics. In particular, in 1536 he published an exposé attacking the corrupt and inadequate practices of Italian doctors. This work, not surprisingly, was detested by the medical community but embraced by the public, and Cardano could be kept from practicing medicine no longer. The College of Physicians in Milan grudgingly accepted him into their ranks in 1539, and soon he shot to the top of his profession. By mid-century, Cardano was perhaps the most famous and sought-after doctor in Europe, one who treated the Pope and even traveled to Scotland—a long and arduous journey in those days—to care for the Archbishop of St. Andrew's.

His days of triumph were not to last, for personal tragedies soon intervened. In 1546, his wife died at age 31, leaving Cardano with two sons and a daughter. Of these, the elder son, Giambattista, was Cardano's hope and joy. The boy proved quite bright, taking his medical degree in Pavia, and appeared to be following his father into a brilliant medical career.

But disaster struck in the form of a "wild woman" (Cardano's words). He related that, on the night of December 20, 1557, " . . . when the desire (to sleep) was about to overcome me, my bed suddenly seemed to tremble, and with it the whole bed-chamber." The next morning, Cardano's inquiries revealed that no other townsperson had felt this nocturnal quake, and Cardano took it as a very bad omen. No sooner had he reached this conclusion than his servant brought the unexpected news that Giambattista had married a woman "utterly without dowry or recommendation."

Indeed, the match proved to be an unfortunate one. Giambattista's wife bore three children, none of whom, she boasted, was Giambattista's. Such infidelity, openly flaunted, brought the young man to the breaking point. In retaliation, he prepared for her a cake laced with arsenic. It did its job all too well, and Giambattista was arrested for murder. Cardano's tireless efforts and great reputation were to no avail; his beloved son was convicted and beheaded in early April 1560.

"This was my supreme, my crowning misfortune," the grieving Cardano wrote. Despondent, he lost his friends, his career, and his zest for life. Moreover, his other son, Aldo, was himself turning into a criminal, and Cardano actually was "obliged to have him imprisoned more than once." Heartbreak seemed to follow heartbreak.

In 1562, he abandoned Milan, the city of his triumphs and tragedies, and accepted a position in medicine at the University of Bologna. With him he took Fazio, Giambattista's son. Between the old man and the boy there developed a strong and loving relationship that perhaps, in his waning years, gave Cardano some of the joy that his own offspring had not.

But the young boy and the new city did not bring tranquility into this stormy life. In 1570, Cardano was arrested and jailed on charges of heresy. At the time, of course, the Church in Italy had adopted a hard line against the unorthodoxies of the Reformation, and it certainly found no comfort in Cardano's casting the horoscope of Jesus or writing the book *In Praise of Nero* about the hated, anti-Christian Roman emperor.

Jailed and humiliated, the aging Cardano seemed to have met with his final disgrace. Yet, thanks to the testimonials of his illustrious friends and the leniency of the Church, Cardano soon got out of prison, went to Rome, and somehow wound up with a pension from the Pope! His was a "Horatio Algebra" story, if ever there was one. Thus resurrected, joined by his beloved grandson, Cardano spent his last years. Although an old man, he noted proudly in his autobiography that he still possessed "fourteen good teeth, and one which is rather weak; but it will last a long time, I think, for it still does its share." Cardano spent his last years in relative tranquility and died quietly, after a very full life, on September 20, 1576.

To the modern reader, Cardano remains a fascinating, if self-contradictory, character. He was incredibly prolific; his collected works fill seven thousand pages and cover a bewildering array of topics, scientific and otherwise. Yet even as he had one foot planted in the modern, rational world, he had another squarely planted in the superstitious irrationality of the Middle Ages. Looking back a century later, the great philosopher and mathematician Gottfried Wilhelm Liebniz summed him up quite aptly: "Cardano was a great man with all his faults; without them, he would have been incomparable."

We now return to the story of the cubic equation, in which Cardano was to play a major role. Recall that, in 1535, Tartaglia of Brescia had soundly bested Antonio Fior by discovering the solution of certain kinds of cubics. Cardano was intrigued. Again and again he wrote to Tartaglia begging for the solution, and again and again he was rebuffed, with Tartaglia vowing to write a book on the matter in his own good time. Initially, Cardano reacted with anger, but eventually more soothing words brought Tartaglia to Milan as Cardano's guest. There, on March 25, 1539, Tartaglia revealed the secret of the depressed cubic—albeit written in cipher—to Cardano, who took the following solemn oath:

> I swear to you by the Sacred Gospel, and on my faith as a gentleman, not only never to publish your discoveries, if you tell them to me, but I also promise and pledge my faith as a true Christian to put them down in cipher so that after my death no one shall be able to understand them.

A final character then appeared in this amazing drama. This was the young Ludovico Ferrari (1522–1565), who arrived at Cardano's door asking for work. Cardano had that very day perceived a good omen in the incessant squawking of a magpie and thus eagerly took the boy in as a servant. It soon became clear that the young Ludovico was extraordinarily precocious. Their relationship quickly turned from master/servant to teacher/pupil and eventually, before Ferrari was 20 years old, to colleague/colleague. Cardano shared Tartaglia's secret with his brilliant young protégé, and together the two of them made astounding progress.

For instance, Cardano discovered how to solve the general cubic equation

$$x^3 + bx^2 + cx + d = 0,$$

where the coefficients b, c, and d may or may not be zero. Unfortunately, Cardano's work rested upon reducing the general cubic to a depressed form and thus ran up against his pledge of secrecy to Tartaglia. Meanwhile, Ferrari succeeded in finding a technique for solving the quartic (or fourth degree) polynomial equation. This was a major discovery in algebra, but it depended upon reducing the quartic to a related cubic, and again Cardano's oath forbade its publication. The two men, possessing the greatest algebraic discoveries of their time, were stymied.

But then, in 1543, Cardano and Ferrari traveled to Bologna where they inspected the papers of Scipione del Ferro, with whom this whole long story had begun nearly three decades earlier. There, in del Ferro's own hand, was the solution to the depressed cubic. To Cardano, the implication was clear: he no longer was prohibited from publishing this result, since it was from del Ferro, and not from Tartaglia, that he would take his cue. The fact that both solutions were identical did not particularly bother the eager Cardano.

And so, in the year 1545, there appeared Cardano's mathematical masterpiece *Ars Magna.* To him, algebra was the "Great Art," and the book represented a breathtaking advance over that which had been previously known. Its 40 chapters begin with simple algebraic matters, but it is in Chapter XI, titled "On the Cube and First Power Equal to the Number," that the world at last saw the solution of the cubic. It is worth noting that Cardano prefaced this key chapter with the following:

> Scipio Ferro of Bologna well-nigh thirty years ago discovered this rule and handed it on to Antonio Maria Fior of Venice, whose contest with Niccolo Tartaglia of Brescia gave Niccolo occasion to discover it. He gave it to me in response to my entreaties, though withholding the demonstration. Armed with this assistance, I sought out its demonstration in [various] forms. This was very difficult.

Cardano had thus given credit where credit was due, which satisfied everyone except Tartaglia. He, on the contrary, raged furiously about Cardano's deceit and treachery. In Tartaglia's eyes, Cardano had violated a sacred oath, pledged on his faith as a "true Christian," and was nothing more nor less than a vile scoundrel. Accusations poured from Tartaglia's pen and were answered not by Cardano, who managed to stay above the fray, but by the tenacious and loyal Ferrari. The latter was known for his hot temper (he had lost a few fingers in an especially vicious fight) and lashed back vehemently. Accusatory, volatile letters flew between Brescia and Milan. For instance, in a 1547 broadside, Ferrari blasted Tartaglia as

> . . . someone who spends the whole time . . . on trifles. I promise you that if it were up to me to reward you, I would load you up so much with roots and radishes that you would never eat anything else in your life.

(The last sentence is a pun on the mathematical roots that permeate cubic problems.)

The conflict culminated in yet another public debate, this one between Tartaglia and Ferrari in Milan on August 10, 1548. Tartaglia later made much of Cardano's absence, blaming him for a cowardly decision "to avoid being present at the dispute." However, the contest, held on Ferrari's home turf, proved a failure for the visitor. Tartaglia blamed this on the rowdiness and partisanship of the crowd, whereas Ferrari naturally attributed the outcome to his own intellectual superiority. In any case, Tartaglia withdrew to return home, and Ferrari was proclaimed the brilliant victor. Mathematics historian Howard Eves, noting the hostile crowd and Ferrari's hot-headed reputation, says that Tartaglia may have been fortunate to escape alive.

These, then, were the events surrounding the solution of the cubic, a story at once complex, lusty, and absurd. It now remains for us to consider the great theorem at the heart of this strange tale.

Great Theorem: The Solution of the Cubic

Upon examining Chapter XI of *Ars Magna*, the modern reader has two surprises in store. One is that Cardano gave not a general proof but a specific example of a depressed cubic, namely

$$x^3 + 6x = 20$$

although in our discussion below we shall treat the more general

$$x^3 + mx = n$$

The second is that his argument was purely geometrical, involving literal cubes and their volumes. Actually, the surprise here is minimized when we recall the primitive state of algebraic symbolism and the exalted position of Greek geometry among Renaissance mathematicians.

The key result of Chapter XI is stated here in Cardano's own words, and his clever dissection of the cube is presented. His wordy "rule" for solving cubics at first sounds quite confusing, but recasting it in a more familiar, algebraic light shows that it does the job.

THEOREM Rule to solve $x^3 + mx = n$:

Cube one-third the coefficient of x; add to it the square of one-half the constant of the equation; and take the square root of the whole. You will duplicate [repeat] this, and to one of the two you add one-half the number you have already squared and from the other you subtract one-half the same . . . Then, subtracting the cube root of the first from the cube root of the second, the remainder which is left is the value of x.

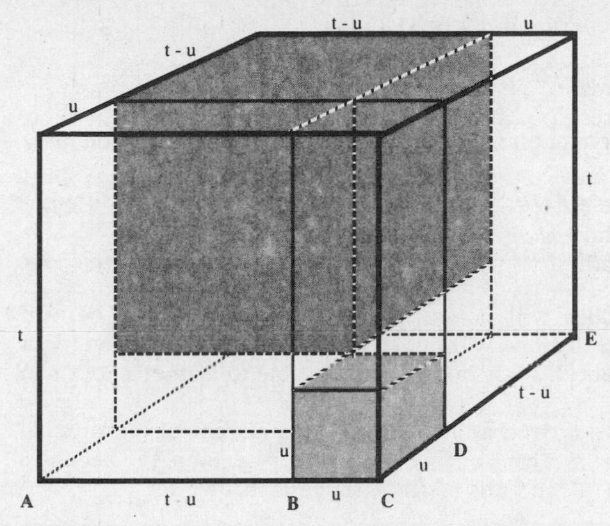

FIGURE 6.1

PROOF Cardano imagined a large cube, having side *AC*, whose length we shall denote by t, as shown in Figure 6.1. Side *AC* is divided at *B* into segment *BC* of length u and segment *AB* of length $t - u$. Here t and u are serving as auxiliary variables whose values we must find. As the diagram suggests, the large cube can be sliced into six pieces, each of whose volumes we now determine:

- a small cube in the lower front corner, with volume u^3
- a larger cube in the upper back corner, with volume $(t - u)^3$
- two upright slabs, one facing front along *AB* and the other facing to the right along *DE*, each with dimensions $t - u$ by u by t (the length of the side of the big cube) and thus each with volume $tu(t - u)$
- a tall block in the upper front corner, standing upon the small cube, with volume $u^2(t - u)$
- a flat block in the lower back corner, beneath the larger cube, with volume $u(t - u)^2$

Clearly the large cube's volume, t^3, equals the sum of these six component volumes. That is,

$$t^3 = u^3 + (t - u)^3 + 2tu(t - u) + u^2(t - u) + u(t - u)^2$$

Some rearrangement of these terms yields

$$(t-u)^3 + [2tu(t-u) + u^2(t-u) + u(t-u)^2] = t^3 - u^3$$

and factoring the common $(t-u)$ from the bracketed expression gives

$$(t-u)^3 + (t-u)[2tu + u^2 + u(t-u)] = t^3 - u^3 \qquad \text{or simply}$$

$$(t-u)^3 + 3tu(t-u) = t^3 - u^3 \qquad (*)$$

(The modern reader will notice that this equation can be derived instantly by simple algebra, without recourse to the arcane geometry of cubes and slabs. But this was not a route available to mathematicians in 1545.)

In (*) we have arrived at an equation reminiscent of the original cubic $x^3 + mx = n$. That is, if we let $t - u = x$, then (*) becomes $x^3 + 3tux = t^3 - u^3$, and this instantly suggests that we set

$$3tu = m \qquad \text{and} \qquad t^3 - u^3 = n$$

If we now can determine the quantities t and u in terms of m and n from the original cubic, then $x = t - u$ will yield the solution we seek.

Ars Magna does not present a derivation of these quantities. Rather, Cardano simply provided the specific rule for solving the "Cube and Cosa Equal to the Number" that was cited above. Trying to decipher his purely verbal recipe is no easy feat and certainly makes one appreciate the concise, direct approach of a modern algebraic formula. Exactly what was Cardano saying in this passage?

To begin, consider his two conditions on t and u, namely

$$3tu = m \qquad \text{and} \qquad t^3 - u^3 = n$$

From the former, we see that $u = m/3t$, and substituting this into the latter yields

$$t^3 - \frac{m^3}{27t^3} = n$$

Multiply both sides by t^3 and rearrange terms to get the equation

$$t^6 - nt^3 - \frac{m^3}{27} = 0$$

At first, this appears to be no improvement whatever, for we have traded our original third-degree equation in x for a sixth-degree equation in t. What saved the day, of course, was that the latter can be regarded as a *quadratic* equation in the variable t^3:

$$(t^3)^2 - n(t^3) - \frac{m^3}{27} = 0$$

The quadratic formula, which had been available to mathematicians for centuries and which we mentioned in the Epilogue to the previous chapter, then yielded:

$$t^3 = \frac{n \pm \sqrt{n^2 + \frac{4m^3}{27}}}{2}$$

$$= \frac{n}{2} \pm \frac{1}{2}\sqrt{n^2 + \frac{4m^3}{27}} = \frac{n}{2} \pm \sqrt{\frac{n^2}{4} + \frac{m^3}{27}}$$

Then, using only the positive square root, we have

$$t = \sqrt[3]{\frac{n}{2} + \sqrt{\frac{n^2}{4} + \frac{m^3}{27}}}$$

Now, we also know that $u^3 = t^3 - n$, and so we conclude that

$$u^3 = \frac{n}{2} + \sqrt{\frac{n^2}{4} + \frac{m^3}{27}} - n \quad \text{or}$$

$$u = \sqrt[3]{-\frac{n}{2} + \sqrt{\frac{n^2}{4} + \frac{m^3}{27}}}$$

At last, we have the algebraic version of Cardano's rule for solving the depressed cubic $x^3 + mx = n$, namely

$$x = t - u$$

$$= \sqrt[3]{\frac{n}{2} + \sqrt{\frac{n^2}{4} + \frac{m^3}{27}}} - \sqrt[3]{-\frac{n}{2} + \sqrt{\frac{n^2}{4} + \frac{m^3}{27}}}$$

Q.E.D.

This expression is called a "solution by radicals" or an "algebraic solution" for the depressed cubic. That is, it involves only the original coefficients in the equation—that is, m and n—and the algebraic operations of addition, subtraction, multiplication, division, and extraction of roots, used only finitely often. A little study shows that this formula yields precisely the same result as Cardano's verbal "Rule" stated above.

Note that the key insight in Cardano's argument was to replace the solution of the cubic by the solution of a related quadratic equation (in t^3). He thus found a way to lower the problem by "one degree" and to move from the unfamiliar turf of cubics to the well-known realm of quadratics. This very clever process suggested a path to follow in attacking equations of the fourth, fifth, and higher degrees well.

As a concrete example, Cardano solved his prototype cubic $x^3 + 6x = 20$. According to his recipe, he first cubed a third of the coefficient of x to get $(\frac{1}{3} \times 6)^3 = 8$; next he squared half of the constant term (that is, half of 20) to get 100, and then added the 8, yielding a sum of 108 whose square root he took. To this he both added and subtracted half of the constant term, to get $10 + \sqrt{108}$ and $-10 + \sqrt{108}$, and finally his solution was the difference of cube roots of these two numbers:

$$x = \sqrt[3]{10 + \sqrt{108}} - \sqrt[3]{-10 + \sqrt{108}}$$

Of course, we could simply substitute $m = 6$ and $n = 20$ into the pertinent algebraic formula. This yields

$$\sqrt{\frac{n^2}{4} + \frac{m^3}{27}} = \sqrt{108} \qquad \text{and so}$$

$$x = \sqrt[3]{10 + \sqrt{108}} - \sqrt[3]{-10 + \sqrt{108}}$$

which is clearly a "solution by radicals." It may come as a surprise—easily checked by a hand calculator—that this sophisticated-looking expression is nothing more than the number "2" in disguise, as Cardano

cubus p̃. 6. rebus æqualis 20.
2. 20.
8. ———— 10.
108.
℞. 108. p̃. 10.
℞. 108. m̃. 10.
℞. v. cu. ℞. 108. p̃. 10.
m̃. ℞. v. cu. ℞. 108. m̃. 10.

Cardano's Rule for the cubic, from *Ars Magna* (photograph courtesy of Johnson Reprint Corporation) correctly pointed out. One readily sees that $x = 2$ is indeed a solution of $x^3 + 6x = 20$.

Further Topics on Solving Equations

Observe that, having found one solution to the cubic, we are now in a position to find any others. For instance, since $x = 2$ solves the specific equation above, we know that $x - 2$ is one factor of $x^3 + 6x - 20$, and long division will generate the other, second-degree factor. In this case, $x^3 + 6x - 20 = (x - 2)(x^2 + 2x + 10)$. The solutions to the original cubic thus arise from solving the linear and quadratic equations

$$x - 2 = 0 \qquad \text{and} \qquad x^2 + 2x + 10 = 0$$

which is easily done. (This particular quadratic has no real solutions, so the cubic has as its only real solution $x = 2$.)

To the modern reader, the next two chapters of *Ars Magna* seem superfluous. Cardano titled Chapter XII "On the Cube Equal to the First Power and Number"—that is, $x^3 = mx + n$—and Chapter XIII was "On the Cube and Number Equal to the First Power"—that is, $x^3 + n = mx$. Today, we would regard these as having already been adequately covered by the formula above, for we would allow m and n to be negative. Mathematicians in the sixteenth century, however, demanded that all coefficients in the equation be positive. In other words, they regarded $x^3 + 6x = 20$ and $x^3 + 20 = 6x$ not just as different equations, but as intrinsically different *kinds* of equations. Such squeamishness about negative numbers is hardly surprising, given Cardano's tendency to think in terms of three-dimensional cubes, where sides of negative length make no sense. Of course, avoiding negatives led to a proliferation of cases and made *Ars Magna* considerably longer than we now find necessary.

So, Cardano could solve the depressed cubic in any of its three versions. But what about the *general* third-degree equation of the form $ax^3 + bx^2 + cx + d = 0$? It was Cardano's great discovery that, by means of a suitable substitution, this equation could be replaced by a related, depressed cubic that was, of course, susceptible to his formula. Before examining this "depressing" process for the cubic, we might take a quick look at it in a more familiar setting—as applied to solving quadratic equations:

Suppose we begin with the general second-degree equation

$$ax^2 + bx + c = 0 \qquad \text{where } a \neq 0$$

To depress it—that is, to eliminate its first-power term—we introduce the new variable y by substituting $x = y - b/2a$ to get

$$a\left(y - \frac{b}{2a}\right)^2 + b\left(y - \frac{b}{2a}\right) + c = 0 \qquad \text{which gives}$$

$$a\left(y^2 - \frac{b}{a}y + \frac{b^2}{4a^2}\right) + by - \frac{b^2}{2a} + c = 0 \qquad \text{or}$$

$$ay^2 - by + \frac{b^2}{4a} + by - \frac{b^2}{2a} + c = 0$$

Then, canceling the by terms, we get the depressed quadratic

$$ay^2 = \frac{b^2}{2a} - \frac{b^2}{4a} - c = \frac{2b^2}{4a} - \frac{b^2}{4a} - \frac{4ac}{4a} = \frac{b^2 - 4ac}{4a}$$

Hence

$$y^2 = \frac{b^2 - 4ac}{4a^2} \qquad \text{and} \qquad y = \frac{\pm\sqrt{b^2 - 4ac}}{2a}$$

Finally

$$x = y - \frac{b}{2a} = \frac{\pm\sqrt{b^2 - 4ac}}{2a} - \frac{b}{2a} = \frac{-b \pm \sqrt{b^2 - 4ac}}{2a}$$

which is of course the quadratic formula once again.

As this example suggests, depressing polynomials can prove quite useful. With this in mind, we return to Cardano's attack on the general cubic. Here, the key substitution is $x = y - b/3a$, which yields

$$a\left(y - \frac{b}{3a}\right)^3 + b\left(y - \frac{b}{3a}\right)^2 + c\left(y - \frac{b}{3a}\right) + d = 0$$

Upon expanding, this becomes

$$\left(ay^3 - by^2 + \frac{b^2}{3a}y - \frac{b^3}{27a^2}\right) + \left(by^2 - \frac{2b^2}{3a}y + \frac{b^3}{9a^2}\right)$$

$$+ \left(cy - \frac{cb}{3a}\right) + d = 0$$

There is but one critical observation we need to make regarding this blizzard of letters, namely, that the y^2 terms will cancel out. Thus, the new cubic loses its second-degree term (as desired). If we divide through by a, the resulting equation takes the form $y^3 + py = q$. We solve this for y by Cardano's formula and from there have no difficulty in determining $x = y - b/3a$.

To see this process in action, consider the cubic

$$2x^3 - 30x^2 + 162x - 350 = 0$$

With the substitution $x = y - b/3a = y - (-30/6) = y + 5$, we get

$$2(y + 5)^3 - 30(y + 5)^2 + 162(y + 5) - 350 = 0$$

which becomes

$$2y^3 + 12y - 40 = 0 \qquad \text{or simply} \qquad y^3 + 6y = 20$$

But this is, of course, the very depressed cubic we solved earlier, and so we know that $y = 2$. Hence $x = y + 5 = 7$, and this checks in the original equation.

Ars Magna did not handle the general cubic quite so concisely as we did here. Instead, demanding only positive coefficients, Cardano had to wade through a string of different cases, such as "On the Cube, Square, and First Power Equal to the Number," "On the Cube Equal to the Square, First Power, and Number," "On the Cube and Number Equal to the Square and First Power," and so on. At last, 13 chapters after solving the depressed cubic, he brought the matter to its conclusion. The cubic had been solved.

Or had it? Although Cardano's formula seemed to be an amazing triumph, it introduced a major mystery. Consider, for instance, the depressed cubic $x^3 - 15x = 4$.

Using $m = -15$ and $n = 4$ in the formula developed above, we get

$$x = \sqrt[3]{2 + \sqrt{-121}} - \sqrt[3]{-2 + \sqrt{-121}}$$

Obviously, if negative numbers were suspect in the 1500s, their *square roots* seemed absolutely preposterous, and it was easy to dismiss this as an unsolvable cubic. Yet it can easily be checked that the cubic above has three different and perfectly real solutions: $x = 4$ and $x = -2 \pm \sqrt{3}$. What was Cardano to make of such a situation—the so-called "irreducible case of the cubic"? He took a few half-hearted stabs at investigating what we now call "imaginary" or "complex" numbers but ultimately dismissed the whole enterprise as being "as subtle as it is useless."

It would be another generation before Rafael Bombelli (ca. 1526–1573), in his 1572 treatise *Algebra*, took the bold step of regarding imaginary numbers as a necessary vehicle that would transport the mathematician from the *real* cubic equation to its *real* solutions; that is, while we begin and end in the familiar domain of real numbers, we seem compelled to move into the unfamiliar world of imaginaries to complete our journey. To mathematicians of the day, this seemed incredibly strange.

We shall examine briefly what Bombelli did. Temporarily disregarding any latent prejudice against $\sqrt{-1}$, we cube the expression $2 + \sqrt{-1}$ to get

$$\begin{aligned}(2 + \sqrt{-1})^3 &= 8 + 12\sqrt{-1} - 6 - \sqrt{-1} \\ &= 2 + 11\sqrt{-1} = 2 + \sqrt{-121}\end{aligned}$$

But if $(2 + \sqrt{-1})^3 = 2 + \sqrt{-121}$, then it surely makes sense to say that

$$\sqrt[3]{2 + \sqrt{-121}} = 2 + \sqrt{-1}$$

Similarly, we can see that $\sqrt[3]{-2 + \sqrt{-121}} = -2 + \sqrt{-1}$. Then, reexamining the cubic $x^3 - 15x = 4$, Bombelli arrived at the solution

$$\begin{aligned}x &= \sqrt[3]{2 + \sqrt{-121}} - \sqrt[3]{-2 + \sqrt{-121}} \\ &= (2 + \sqrt{-1}) - (-2 + \sqrt{-1}) = 4\end{aligned}$$

which is correct!

Admittedly, Bombelli's technique raised more questions than it resolved. For one thing, how does one know beforehand that $2 + \sqrt{-1}$ is going to be the cube root of $2 + \sqrt{-121}$? It would not be until the middle of the eighteenth century that Leonhard Euler could give a sure-fire technique for finding roots of complex numbers. Furthermore, what exactly were these imaginary numbers, and did they behave like their real cousins?

It is true that the full importance of complex numbers did not become evident until the work of Euler, Gauss, and Cauchy more than two centuries later, and we shall meet this topic again in the Epilogue to Chapter 10. Still, Bombelli deserves credit for recognizing that such numbers have a role to play in algebra, and he thereby stands as the last in the line of the great Italian algebraists of the sixteenth century.

One point should be stressed here. Contrary to popular belief, imaginary numbers entered the realm of mathematics not as a tool for solving *quadratics* but as a tool for solving *cubics*. Indeed, mathematicians could easily dismiss $\sqrt{-121}$ when it appeared as a solution to $x^2 + 121 = 0$ (for this equation clearly has no real solutions). But they could not so easily ignore $\sqrt{-121}$ when it played such a pivotal role in yielding the solution $x = 4$ for the previous cubic. So it was cubics, not quadratics, that gave complex numbers their initial impetus and their now-undisputed legitimacy.

We should make a final observation about *Ars Magna*. In Chapter XXXIX, Cardano introduced the solution of the quartic with the words:

> There is another rule, more noble than the preceding. It is Lodovico Ferrari's, who gave it to me on my request. Through it we have all the solutions for equations of the fourth power.

While the procedure is quite complicated, its two key steps should ring a bell:

1. Beginning with a general quartic $ax^4 + bx^3 + cx^2 + dx + e = 0$, depress it using the substitution $x = y - b/4a$ and then divide through by a, to generate a depressed quartic in y:

$$y^4 + my^2 + ny = p$$

2. By cleverly introducing auxiliary variables, replace this quartic by a related *cubic*, which then can be solved using the techniques developed above. Here again, Ferrari invoked the rule-of-thumb that the way to solve an equation of a given degree is to reduce it to the solution of an equation of one degree less.

Those who were capable of reading through this, and all of the other discoveries in *Ars Magna*, must have been breathless by the time they finished. The art of equation solving had been taken to new heights, and Luca Pacioli's original assessment that cubics, let alone quartics, were beyond the reach of algebra had been shattered. It is little wonder that Cardano ended his book with the enthusiastic and rather touching statement: "Written in five years, may it last as many thousands."

Epilogue

One question that the Cardano-Ferrari work left unanswered was the algebraic solution of the quintic, or fifth-degree, equation. Their efforts certainly suggested that such a solution by radicals was possible and even gave an obvious hint as to how to begin. That is, faced with the quintic

$$ax^5 + bx^4 + cx^3 + dx^2 + ex + f = 0$$

introduce the transformation $x = y - b/5a$ to depress it to

$$y^5 + my^3 + ny^2 + py + q = 0$$

and then search for some auxiliary variables to reduce this to a quartic equation, which is known to be solvable by radicals. Such an argument was especially appealing not only because it mimicked the approach that had proved so successful in disposing of cubic and quartic equations, but also because, as was well known, any fifth-degree (or, indeed, any odd-degree) polynomial equation *must* have at least one real solution. This follows because the graphs of odd-degree equations look something like that of the specific fifth-degree equation shown in Figure 6.2. That is, they rise ever higher as we move in one direction along the x-axis and fall ever lower as we move in the other direction. Consequently, such functions must be positive somewhere and must be negative somewhere else, and we conclude—using a result technically known as the intermediate value theorem—that the continuous graph must somewhere cross the x-axis. In the diagram of the quintic above, c is such a point, and hence $x = c$ is a solution to $x^5 - 4x^3 - x^2 + 4x - 2 = 0$. A

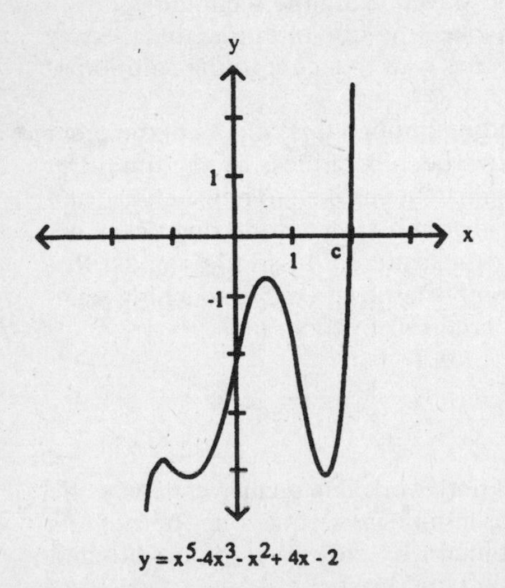

FIGURE 6.2

similar argument guarantees that any odd-degree polynomial equation has (at least) one real solution.

Note, however, that although the intermediate value theorem says that real solutions for quintics *exist*, it by no means gives them explicitly. It was the precise formula for such solutions that the algebraists who followed Ferrari were seeking.

Alas, all efforts in this direction—and they were numerous—met with failure. A century passed, and another, yet no one could provide a "solution by radicals" for the quintic. This came in spite of the fact that later mathematicians found a transformation to reduce the general quintic to one of the form

$$z^5 + pz = q$$

If we called the earlier equation "depressed," this one must have been "utterly despondent." Yet even this highly simplified quintic resisted the efforts of all who attacked it. The situation was frustrating, if not slightly scandalous.

Then, in 1824, a young Norwegian mathematician, Niels Abel (1802–1829), shocked the mathematical world by showing that no "solution by radicals" was possible for fifth- or higher degree equations. The search, in short, had been doomed from the start. Abel's proof, which can be found in D. E. Smith's *Source Book in Mathematics*, is quite advanced and not at all easy to follow, yet it certainly stands as a landmark in mathematics history.

It is worth noting what Abel's result did and did not imply. He did not say that *no* quintic is solvable, for we obviously can get lucky and solve such equations as $x^5 - 32 = 0$, which clearly has the solution $x = 2$. Further, Abel did not deny that we might solve quintics using techniques other than the algebraic ones of adding, subtracting, multiplying, dividing, and extracting roots. Indeed, the general quintic *can* be solved by introducing entities called "elliptic functions," but these require operations considerably more complicated than those of elementary algebra. In addition, Abel's result did not preclude our approximating solutions for quintic equations as accurately as we—or our computers—wish.

What Abel did do was prove that there exists no algebraic formula, involving only the coefficients of the original quintic equation, that will be a guaranteed generator of solutions. The analogue of the quadratic formula for second-degree equations and Cardano's formula for cubics simply does not exist—it is impossible to provide a universally effective means of finding solutions by radicals for quintics.

The situation is reminiscent of that encountered when trying to square the circle, for in both cases mathematicians are limited by the tools they can employ. For circle squaring, as noted in Chapter 1, the compass and straightedge are simply not powerful enough to get the job done. Likewise, it is the restriction to "solutions by radicals" that hampers mathematicians in their pursuit of the quintic. The familiar operations of algebra are incapable of taming something as wild as a fifth-degree equation.

We seem to be on the brink of a paradox here, for although mathematicians know that quintics must have solutions, Abel showed that there is no algebraic way of finding them. But it is that modifier, algebraic, that keeps us from plunging over the brink into mathematical chaos. Indeed, what Abel actually demonstrated was that algebra does have very definite limits, and for no obvious reason, these limits appear precisely as we move from the fourth to the fifth degree.

Consequently, in a very real sense, we have come full circle. The pessimism of Luca Pacioli, obscured by the thrill of discovery in the sixteenth century, turned out to have been prophetic. When we move beyond fourth-degree equations, the unequivocal triumph of algebra is lost forever.

Chapter 7

A Gem from Isaac Newton

(Late 1660s)

Mathematics of the Heroic Century

If the sixteenth century saw a quickened pace of mathematical activity, the seventeenth would bring an absolute onslaught of innovation and discovery. It has come to be called the heroic century in the history of mathematics because of the intellectual giants who came and went during these productive years.

At this time, the focus of activity shifted northward, from the talented Italian algebraists of the previous chapter to an array of French, German, and British thinkers. The causes of the shift were certainly many and, as with any human endeavor, involved a healthy dose of pure chance. But some scholars, in noting this phenomenon, have attached importance to the relative freedom of inquiry in northern Europe as contrasted with the harsh restrictions imposed by the Church in Italy. The fate of Galileo is the best-known case in point, the story of a scientist whose investigations led him into topics unacceptable to the powerful religious authorities of seventeenth-century Roman Catholicism. Galileo's imprisonment and forced disavowal would surely have chilled the intellectual climate, and the entire episode constitutes one of the most ignoble chapters in the history of science.

All was not freedom and openness in the north, yet the heritage of the Reformation seemed to favor the kind of unfettered inquiry that would loosen the intellect of a Kepler or Descartes or Newton. It is possible that, in attempting to impose an inflexible orthodoxy, the Church had condemned Italy to a second-class citizenship in science.

It was not mathematics alone that was flourishing as the sixteenth century became the seventeenth. In 1607, the British settled at Jamestown, and the European colonization of the New World began in earnest. A few years before Jamestown, Galileo had investigated the motion of falling bodies with such care and ingenuity that he forever altered the nature of physics. Two years after Jamestown, the same Galileo turned the recently invented "spyglass" toward the heavens and launched modern astronomy, even as he simultaneously began the journey that would lead to the personal crisis already mentioned. And, not to overlook the arts, we should recall that Cervantes wrote his monumental *Don Quixote* in 1605, while in 1601 an English playwright named William Shakespeare had written a play he titled *Hamlet, Prince of Denmark.*

Obviously, cultural landmarks do not fit themselves neatly into 100-year intervals, and it was in the latter years of the previous century that the first signs appeared of the mathematical revolution that was in store. A "heroic century" needs heroes, and we shall take a brief look at some of them.

In the 1590s, the French mathematician Francois Viète published his influential *In artem analyticam isagoge* (usual translation: *The Analytic Art*). Viète's approximation of π was mentioned in Chapter 4, but it is this 1591 work that stands as his masterpiece. *In artem* went a long way toward developing symbolic algebra, destined to become the "alphabet" of higher mathematics. Admittedly, Viète's algebraic notation looks far from modern and seems unduly cumbersome and wordy for those accustomed to today's mathematics. For instance, Viète would write

D in R − D in E aequabitur A quad

as his version of our modern $DR - DE = A^2$. Yet he made an important step toward the use of letters to designate quantities in an equation. Refined and extended in the decades ahead, algebraic symbolism would transform the look—and the substance—of mathematics in the new century.

Early in the 1600s, a pair of mathematicians from the British Isles, John Napier (1550–1617) and Henry Briggs (1561–1631), jointly introduced, perfected, and exploited the "logarithm," a concept having tremendous practical and theoretical significance. Logarithms have the remarkable property of simplifying such otherwise tedious computations as multiplication, division, and the extraction of roots so that no scientist of sound mind would thereafter go about finding $\sqrt[7]{234.65}$ without the benefit of logarithms. In the next century, Pierre-Simon Laplace would observe that the logarithms of Napier and Briggs "by shortening the labors doubled the life of the astronomer." Certainly, the cooperation between Briggs and Napier in this venture is noteworthy and stands in striking contrast to some of the bitter disputes and jealousies that would plague mathematics in the years to come.

As the century progressed, three French mathematicians commanded the spotlight. One was René Descartes (1596–1650), a philosopher and mathematician whose 1637 *Discours de la méthode* became a landmark in the history of philosophy. It was a treatise on "universal science" that both anticipated and spurred on the great scientific explosion that characterized the times. While the philosophical content of the *Discours* was widely discussed and hotly debated, it was an appendix, titled *La géométrie*, that most directly influenced the development of mathematics. There Descartes provided the first published account of what we now call analytic geometry. Like Viète's algebraic notation, Descartes' analytic geometry was far from modern in appearance, but it announced the marriage of algebra and geometry that would become indispensable in all subsequent mathematical work.

When the *Discours* appeared, Blaise Pascal (1623–1662), although a youngster of 14, was already attending meetings of senior French mathematicians. He was on the brink of a spectacular, if short, mathematical career. Pascal was a brilliant child, of a kind one sometimes encounters in the history of mathematics. At age 16, he had so impressed the mighty Descartes with his mathematical writings that the latter refused to believe that so young a boy was the author. Two years later, Pascal invented a calculating machine that stands as the very remote predecessor of the modern computer. Further, he made significant contributions to the theory of probability, thereby pushing the subject beyond the rudiments provided by Cardano a century before.

In spite of his obvious talent for mathematics, Pascal devoted most of his adult life to questions of theology, and his work in this area is still regularly studied. A man who often perceived omens in events around him, Pascal concluded that God's plan for him did not include mathematics and dropped the subject entirely. However, while experiencing a particularly nagging toothache when he was 35, Pascal let his thoughts wander to mathematics, and the pain disappeared. He took this as a heavenly sign and made a quick but intensive return to mathematical research. Although this lasted barely a week, Pascal managed to discover the fundamental properties of the cycloid curve, a topic discussed in the next chapter. With that, Pascal again abandoned mathematics, and in 1662, at the age of 39, he died.

This brings us to perhaps the most remarkable of the triumvirate of Frenchmen who dominated the mathematical landscape in the mid-seventeenth century, Pierre de Fermat (1601–1665) of Toulouse. Fermat is notable for the many areas of mathematics in which he made significant discoveries. He created his own analytic geometry independent of—and perhaps earlier than—Descartes, and in some ways Fermat's approach to the topic was more "modern" than that of his famous contemporary. Of course, Descartes was the first to publish a description of analytic geometry and thereby reaped more of the glory, but Fermat deserves equal credit for its creation. Likewise, a lively correspondence between Pascal and Fermat laid the previously mentioned foundation of probability theory in the 1650s. As if that were not enough, Fermat made great strides toward developing what we now call differential calculus. In some places, France in particular, he is sometimes accorded the honor of being a co-founder of the calculus, although most historians of mathematics, while recognizing his achievements, feel that this is going a bit too far.

But it was in the area of number theory that Fermat left his most indelible mark. We have seen this topic before, with the writings of Euclid in Books VII–IX of the *Elements*. One classic work on this subject was the *Arithmetica* of Diophantus (ca. A.D. 250?). When this was rediscovered and translated during the Renaissance, it proved to be a very influential treatise. Fermat acquired a copy, devoured Diophantus' writings, and soon was making profound discoveries of his own about the amazing properties of the whole numbers.

It was common for him to state an intriguing result, and sometimes claim to have a valid proof, but rarely did he actually provide it. Thus it was the job of later mathematicians—and more often than not, this meant Euler—to furnish the missing verifications. As a result, mathematics historians are left with the dilemma of determining to whom the credit belonged—to Fermat, who had first stated the result and who may well have had a proof, or to Euler, who actually wrote down such an argument.

By far the most intriguing of Fermat's "theorems" (the term is used gingerly since so many of his results were stated with an excess of confidence but an absence of proof) was triggered by the work of Diophantus. In his personal copy of the *Arithmetica*, Fermat scribbled a note in the margin beside Proposition II.8, a result about expressing a perfect square as the sum of two other perfect squares. Instances of such a decomposition abound, as in $5^2 = 3^2 + 4^2$ or $25^2 = 7^2 + 24^2$. Next to Diophantus' theorem, Fermat jotted these famous lines:

> But it is impossible to divide a cube into two cubes, or a fourth power [*quadratoquadratum*] into two fourth powers, or generally any power beyond the square into two like powers; of this I have found a remarkable demonstration. This margin is too narrow to contain it.

In modern parlance, his jottings claimed that we can never find whole numbers a, b, and c and an exponent $n \geq 3$ for which $a^n + b^n = c^n$. If he were correct, then the decomposition of a perfect square into the sum of two perfect squares was really something of a fluke; except for squares, said Fermat, numbers of a given power cannot be written as the sum of two smaller numbers of that same power.

As usual, there was no proof. Fermat attributed this omission merely to the narrow margins on the page of the Diophantus text. If only he had had a blank page to work with, Fermat seemed to say, he would gladly have provided the wonderful demonstration he claimed to have discovered. Instead, as with most of his assertions, the task of finding a proof was left to posterity.

In this case posterity is still looking, for Fermat's assertion remains to this day unresolved. Even Euler, who unraveled the mysteries of so many of Fermat's "theorems," could prove this assertion only for $n = 3$ and $n = 4$. That is, Euler showed that indeed a cube cannot be written as the sum of two cubes, nor a fourth power as the sum of two fourth powers. But resolution of the general case, today popularly known as "Fermat's last theorem," is still anybody's guess. The betting odds seem to be that Fermat was correct in this, as in so many other of his unproved assertions. Nonetheless, no number theorist to date has managed to prove the result true, nor has anyone cooked up a counterexample to show it false. In this sense, then, calling it his last "theorem" is surely a bit premature. Should anyone resolve the issue—and even in the late twentieth century interest in the problem runs high—he or she will surely merit a page in all subsequent histories of mathematics.

And so, if we were somehow to return to the summer of 1661 and survey the mathematical legacy of the seventeenth century, we would have much of significance to behold. Algebraic notation, logarithms, analytic geometry, probability, and the theory of numbers—all had come into their own, and the names of Viète, Napier, Descartes, Pascal, and Fermat would be justly revered. They were heroes indeed. Of course, no one on that summer's day would have paid the least attention to the quiet beginning of a mathematical journey that would soon eclipse them all. The journey was to commence at the beautiful Trinity College of Cambridge University. In the summer of 1661, a young lad from nearby Woolsthorpe was about to begin his university career. He had previously shown promise, but much the same might have been said about the dozen or so equally obscure classmates with whom he began his Trinity experience. But *this* lad was on his way to become the supreme hero of the heroic century and, simultaneously, to change the way mankind would ever after look at the world. His name, of course, was Isaac Newton.

A Mind Unleashed

Newton was born on Christmas Day of 1642, a dangerously premature infant frail and tiny enough to be put into "a quart pot." Adding to his woes, his father had died in early October, so Newton's mother was left alone to care for the delicate newborn. Somehow, the baby overcame these initial dangers and the harsh Lincolnshire winter, and in the end Isaac would live to the impressive age of 84.

Although the physical hardships were overcome, those of a personal nature proved far more lasting. When Newton was three, his mother, Hannah Ayscough Newton, married Barnabas Smith, the 63-year-old rector of a nearby village. Smith, although eager to have a young wife, was not in the market for a 3-year-old as well. Thus, when Newton's mother moved in with her new husband, young Isaac was left behind to be cared for by his grandmother. The pain of separation from his only surviving parent must have been overwhelming. Her proximity would have been sheer torture, for he could easily have climbed a tree and gazed out across the heath to see the village church spire where his mother and stepfather lived. Isaac, a child who had never known his father, had now lost his mother as well, not to the scourge of disease but to the cruelty of indifference. As we shall see, Newton would grow into a neurotic and misanthropic adult, one who rarely experienced the warm glow of human friendship. It is an easy matter to attribute such character traits to his abandonment by the one person about whom his world had revolved.

As Isaac grew, he received a respectable grammar school education in the style of the day, which is to say an education heavy on the study of Latin and Greek. Outside the school, he kept largely to himself, occupying his time by reading or building various miniature devices of great charm and delicacy. It is said that he built a small working windmill driven by a mouse upon a treadmill; he made sundials and positioned them at various key points about his quarters; or he took to attaching a lighted lantern to a high-flying kite on dark spring evenings, a phenomenon that must have terrorized residents of the quiet English countryside. Such activities indicate a very agile young mind not totally absorbed by the conjugation of Latin verbs. They also presage the gifted experimental physicist whose practical laboratory devices would prove invaluable in the development of his theories.

And so it was that Isaac Newton set off for Trinity College, Cambridge, in the summer of 1661. At that time, the quiet town on the River Cam had already been a site of higher learning for 400 years, so it was an old and established institution in which Newton found himself. Cambridge had especially flourished earlier in the century, with the rise of Puritanism and Reformation zeal in England. It could boast everything from the King James translation of the Bible, to the architectural masterpiece that was King's College Chapel, to that most Puritan of leaders, Oliver Cromwell, who hailed from nearby Huntingdon and had attended Sidney Sussex College of Cambridge until 1617.

When Newton arrived, much of this past glory was in jeopardy, and the reasons were intimately linked to the vicissitudes of recent British history. In 1642, the year of Newton's birth, the Puritans under Cromwell had brought to a successful conclusion their long campaign against the monarchy. Cromwell himself assumed control of the English government, an authority that became unquestioned when, in 1649, King Charles I was beheaded in Whitehall, London. The Royalists, whose base of support was anchored at Oxford University, were in temporary eclipse while the Puritans of Cambridge had their days of glory.

These days were not to last, however. The Puritan Commonwealth proved to be little better, and perhaps even worse, than the monarchy it had replaced. When Cromwell died in 1658, no Puritan leader could fill the void, and British sentiment soon called for a return to kingly rule. Thus, in 1660 Charles II, son of the beheaded king, was placed upon the throne in what came to be known as the Restoration. Needless to say, this turned the tables. Cambridge University became a natural target for the suspicions and hostility of the newly empowered Royalists. When Newton arrived the year after the Restoration, the place was characterized by political intrigue, patronage, and lethargy. It was far from an ideal situation.

We who today revere Cambridge as one of the handful of truly great centers of learning may find hard to believe the state to which conditions had decayed in the 1660s. Professors were appointed for political or ecclesiastical reasons, and, for many, scholarship was simply irrelevant. There are records of faculty members who occupied their positions for half a century without having a single student, or writing a single book, or giving a single lecture! Some, in fact, did not even live in the Cambridge area, and visited only infrequently.

With such indifference among the professoriate, it is no surprise that the students were often equally immune to the pursuit of higher learning. On paper, the university maintained the pretense of a vibrant intellectual life, heavily slanted toward the classical curriculum, for its scholarly young men. In reality, Cambridge students indulged rather heavily in such activities as drinking at the ubiquitous pubs. It was certainly possible, for student and professor alike, to coast through Cambridge with very little intellectual exertion.

At first, Isaac Newton took the stated expectations at face value. He began the prescribed courses in Latin literature and Aristotelian philosophy but gradually abandoned the project, either because he realized that he knew more than his tutors, or because he recognized the irrelevance of the moribund curriculum, or simply because it was clear that no one really cared whether or not he did the work.

His colleagues at Trinity probably had the same reaction, but whereas they were likely to run off to a pub for their nightly carousing, Newton turned his attention elsewhere. He read voraciously and could be seen walking across the grounds in deep contemplation. When an idea had captured his interest, Newton could be impossibly single-minded and would often neglect to eat or sleep in favor of long bouts with an especially intriguing problem. He also displayed, especially early in his Cambridge days, a streak of old-fashioned guilt, evident in the fact that he kept a list of his sins in a notebook. These included everything from a failure to pray often enough, to being inattentive at church services, to "having uncleane thoughts, words, actions, and dreamese." Certainly the Puritan strain ran deep, but one might expect as much from a solitary, introspective young man who grew up where and when he did.

When not sinning or writing about it, this ever-curious student would conduct experiments on the nature of light, color, and vision. For instance, he once stared at the sun for an excessively long period, then dutifully recorded the spots and flashes that affected his vision for days afterward; in fact, he had to confine himself to the darkness for some time to let the images gradually fade. On another occasion, curious about the effect of the eyeball's shape in distorting and altering vision, he devised a particularly gruesome experiment with himself as subject. As Newton described it, he took a small stick, or "bodkin," and pushed it

> betwixt my eye and ye bone as neare to ye backside of my eye as I could, and pressing my eye with ye end of it . . . there appeared severall white,

> darke, and coloured circles, which circles were plainest when I continued to rub my eye with a point of ye bodkin. . . .

This disgusting procedure was illustrated by a drawing in Newton's hand showing the stick sliding under and behind his distorted eyeball, nicely labeled with letters from *a* to *g*. Clearly, this was no ordinary undergraduate.

For all the deficiencies of Restoration Cambridge, it did possess a wonderful library, the very resource necessary for an inquiring mind of the first order. There is a story that, while at Sturbridge Fair in 1663, Newton picked up a book on astrology. To help him understand its geometric diagrams, he decided he needed to acquaint himself with Euclid's *Elements*. Interestingly, on his first reading he found this ancient text to be full of trivial and self-evident results (an opinion, by the way, which a mature Newton would abandon).

One striking thing about Newton's readings at this time was that he was not content to stick with the Greek classics. He devoured Descartes' geometry, but only with great effort. He later recalled beginning the work and reading a few pages until utterly stumped. Then he would return to page 1 and begin anew, this time penetrating a bit further before the writings became incomprehensible. Again he would start over, and by this gradual process he plowed his way through *La géométrie*, without the assistance of a single tutor or professor. Of course, given the sorry state of affairs among the staff and the classical flavor of the stated curriculum, he may have had difficulty finding anyone qualified to help.

There was one Cambridge professor, however, who indeed had such qualifications. He was Isaac Barrow (1630–1677), occupant of the prestigious Lucasian Chair of Mathematics. While Barrow was certainly not Newton's teacher in anything like a modern sense, he undoubtedly had some contact with the budding scholar and may have directed Newton to the major sources of contemporary mathematics. By reading and thinking constantly, Newton advanced from a fairly ordinary scientific and mathematical background to a mastery of the most up-to-date discoveries of the time. Having brought himself to the frontier, he was now ready to move into uncharted territory.

In 1664, Newton was promoted to the status of scholar at Trinity, thus acquiring a four-year period of financial support toward a master's degree. This promotion brought with it even greater freedom to follow his interests, and this freedom, combined with the solid background of his readings, was about to unleash one of the most extraordinary intellects in history. More than ever, Newton applied his intense, almost unbelievable, powers of concentration to attack the problems before him. In the twentieth century, the famous Cambridge economist John Maynard Keynes made this assessment of Newton's abilities:

> His peculiar gift was the power of holding continuously in his mind a purely mental problem until he had seen straight through it. . . . Anyone who has ever attempted pure scientific or philosophical thought knows how one can hold a problem momentarily in one's mind and apply all one's powers of concentration to piercing through it, and how it will dissolve and escape and you find that what you are surveying is a blank. I believe that Newton could hold a problem in his mind for hours and days and weeks until it surrendered to him its secret.

Newton's own explanation of how he solved his wonderful problems was similar, if a bit more prosaic: "by thinking on them continuously."

In the feverish excitement of discovery, the next few years would see him ever more often working late into the night by the glimmer of candlelight, and it is said that his uncomplaining cat grew fat eating Newton's untouched dinners. But missed meals and lost sleep were inconsequential compared to the strides this young man was making.

He had embarked upon what were probably the most productive two years that any thinker—certainly any 23-year-old thinker—has ever experienced. His days of triumph were spent in part at Cambridge and in part back at Woolsthorpe because of the closing of the university necessitated by an outbreak of the dreaded plague. Early in 1665, he discovered what we now call the "generalized binomial theorem," which became a major component of his subsequent mathematical works. Soon thereafter he came upon his "method of fluxions"—today better known by the name of differential calculus—and in 1666 had devised the "inverse method of fluxions"—that is, the integral calculus. In between, he formulated his groundbreaking theory of colors. But Newton recalled that there was more:

> . . . the same year I began to think of gravity extending to ye orb of the Moon & . . . I deduced that the forces which keep the Planets in their Orbs must [be] reciprocally as the squares of their distances from the centers about which they revolve: & thereby compared the force requisite to keep the Moon in her Orb with the force of gravity at the surface of the earth, & found them to answer pretty nearly.

These recollections, made half a century later by the aged Newton, described perfectly the embryonic theory of universal gravitation upon which, more than any other single achievement, rests his scientific fame. Surveying these discoveries, he observed with a remarkable candor and nonchalance:

> All this was in the two plague years of 1665–1666. For in those days I was in the prime of my age for invention & minded Mathematicks & Philosophy more then at any time since.

These two plague years have since come to be called Newton's *anni mirabiles*, or "wonderful years," and indeed they were. A popular legend says that during this time all of his theories emerged, complete and full-blown, from his fertile mind. This is surely an exaggeration, for in the years ahead Newton would continually refine and improve upon these theories. Yet the burst of creativity he exhibited in this short span of time did define and direct not only the research of his own lifetime but in a very substantial way the future of science itself.

Today, it is easy to forget that Newton made these extraordinary discoveries as a totally anonymous Cambridge student. Professor R. S. Westfall, perhaps the foremost Newton biographer of our day, addressed this remarkable fact in the following memorable passage:

> [Newton's triumph] was a virtuoso performance that would have left the mathematicians of Europe breathless in admiration, envy, and awe. As it happened, only one other mathematician in Europe, Isaac Barrow, even knew that Newton existed, and it is unlikely that in 1666 Barrow had any inkling of his accomplishment. The fact that he was unknown does not alter the other fact that the young man not yet twenty-four, without benefit of formal instruction, had become the leading mathematician of Europe. And the only one who really mattered, Newton himself, understood his position clearly enough. He had studied the acknowledged masters. He knew the limits they could not surpass. He had outstripped them all, and by far.

As we have seen, the focus of mathematical scholarship has shifted from place to place throughout history, from the Pythagoreans at Crotona, to Plato's Athenian Academy, to Alexandria, to Baghdad, and then to the Renaissance Italy of Cardano and Ferrari. Incredibly, in the mid-1660s, it came to rest in the modest rooms of a Trinity College student, and wherever Newton was, there too was the mathematical center of the world.

Newton's Binomial Theorem

From his remarkable output, we can here examine only a tiny, tiny fraction. We shall start with the binomial theorem, Newton's first great mathematical discovery. Paradoxically, it was not a "theorem" in the sense of Euclid or Archimedes in that Newton did not furnish a complete proof. Yet his insight and intuition served him well enough to devise the germane formula and, as we shall see, apply it in the most wonderful fashion.

The binomial theorem dealt with expanding expressions of the form $(a + b)^n$. With only simple algebra and adequate perseverance one obtains such formulas as

$$(a + b)^2 = a^2 + 2ab + b^2$$
$$(a + b)^3 = a^3 + 3a^2b + 3ab^2 + b^3$$
$$(a + b)^4 = a^4 + 4a^3b + 6a^2b^2 + 4ab^3 + b^4$$

and so on. Obviously it would be desirable to be able to find the coefficient of a^7b^5 in the expansion of $(a + b)^{12}$ without going through the tedious calculations of actually multiplying $(a + b)$ by itself a dozen times. The question of expanding the binomial had been raised, and solved, long before Newton was born. The Chinese mathematician Yang Hui knew the secret in the thirteenth century, although his work was unknown in Europe until relatively recent times. Viète likewise ran through binomial powers in Proposition XI of the Preliminary Notes of *In artem*. But it was Blaise Pascal who got his name attached to the great discovery. Pascal noted that the coefficients could be easily obtained from the array now known as "Pascal's triangle":

1
1 1
1 2 1
1 3 3 1
1 4 6 4 1
1 5 10 10 5 1
1 6 15 20 15 6 1
1 7 21 35 35 21 7 1
and so on

where each entry in the body of the triangle is obtained by adding the numbers in the row above to the left and right. Thus, according to Pascal, the next row would be

1 8 28 56 70 56 28 8 1

Note, for example, that the entry 56 arises from adding 21 + 35, the numbers to the left and right in the previous row.

The link between Pascal's triangle and the expansion of $(a + b)^8$ is immediate, since the last line of the triangle gives us the needed coefficients. That is,

$$(a + b)^8 = a^8 + 8a^7b + 28a^6b^2 + 56a^5b^3 + 70a^4b^4 + 56a^3b^5 + 28a^2b^6 + 8ab^7 + b^8$$

By extending the triangle a few more lines, we come upon 792 as the coefficient of a^7b^5 in the expansion of $(a + b)^{12}$. The utility of the triangle is thus quite evident.

The young Newton, when his thinking turned to the expansion of the binomial, was able to devise a formula for generating the binomial coefficients directly, without the tedious process of constructing the triangle down to the necessary row. Further, his inherent belief in the persistence of patterns suggested to him that the formula that correctly generated coefficients for binomial powers like $(a + b)^2$ or $(a + b)^3$ should work just as well for powers like $(a + b)^{1/2}$ or $(a + b)^{-3}$.

Here a word about fractional and negative exponents is needed. In an elementary algebra course, we learn that $a^{1/n} = \sqrt[n]{a}$ while $a^{-n} = 1/a^n$. While Newton may not have been the first to recognize these relationships, he certainly made the most of them in conjunction with the binomial expansion of expressions like $\sqrt{1 + x}$ or $1/(1 - x^2)$.

Newton's version of his binomial expansion is presented here as he explained it in a significant 1676 letter to his great contemporary Gottfried Wilhelm Leibniz (a letter delivered via Henry Oldenberg of the Royal Society). Newton wrote:

$$(P + PQ)^{m/n} = P^{m/n} + \frac{m}{n}AQ + \frac{m - n}{2n}BQ + \frac{m - 2n}{3n}CQ + \frac{m - 3n}{4n}DQ + \cdots$$

where $P + PQ$ is the binomial to be considered; where m/n is the power to which we shall raise the binomial "whether that power is integral or (so to speak) fractional, whether positive or negative"; and where A, B, C, and so on represent the immediately preceding terms in the expansion.

For those who have seen the binomial expansion in modern guise, Newton's statement here may look perplexing and unfamiliar. But a closer examination should resolve any questions. That is, we first notice that

$$A = P^{m/n}$$

$$B = \frac{m}{n} AQ = \frac{m}{n} P^{m/n} Q$$

$$C = \frac{m-n}{2n} BQ = \frac{(m-n)m}{(2n)n} P^{m/n} Q^2 = \frac{\left(\frac{m}{n}\right)\left(\frac{m}{n}-1\right)}{2} P^{m/n} Q^2$$

$$D = \frac{m-2n}{3n} CQ = \frac{\left(\frac{m}{n}\right)\left(\frac{m}{n}-1\right)\left(\frac{m}{n}-2\right)}{3 \times 2} P^{m/n} Q^3 \qquad \text{and so on}$$

Then, applying Newton's formula and factoring the common $P^{m/n}$ from both sides of the equation, we arrive at

$$P^{m/n}(1+Q)^{m/n} = (P+PQ)^{m/n} = P^{m/n}\left[1 + \frac{m}{n}Q + \frac{\left(\frac{m}{n}\right)\left(\frac{m}{n}-1\right)}{2} Q^2 + \frac{\left(\frac{m}{n}\right)\left(\frac{m}{n}-1\right)\left(\frac{m}{n}-2\right)}{3 \times 2} Q^3 + \cdots\right]$$

Upon canceling $P^{m/n}$, what remains is

$$(1+Q)^{m/n} = 1 + \frac{m}{n}Q + \frac{\left(\frac{m}{n}\right)\left(\frac{m}{n}-1\right)}{2} Q^2 + \frac{\left(\frac{m}{n}\right)\left(\frac{m}{n}-1\right)\left(\frac{m}{n}-2\right)}{3 \times 2} Q^3 + \cdots$$

which may look a bit more familiar.

We would do well to follow Newton's lead and use this in a few specific examples. For instance, in expanding $(1+x)^3$, we replace Q by x and m/n by 3 to get

$$\begin{aligned}(1+x)^3 &= 1 + 3x + \frac{3\times 2}{2}x^2 + \frac{3\times 2\times 1}{3\times 2}x^3 + \frac{3\times 2\times 1\times 0}{4\times 3\times 2}x^4 + \cdots \\ &= 1 + 3x + \frac{6}{2}x^2 + \frac{6}{6}x^3 + \frac{0}{24}x^4 + \frac{0}{120}x^5 + \cdots \\ &= 1 + 3x + 3x^2 + x^3\end{aligned}$$

Note that this is precisely the pattern generated by Pascal's triangle; moreover, since our original exponent was the positive integer 3, the expansion terminated after four terms.

Quite a different phenomenon awaited Newton when the exponent was negative. As an example, in expanding $(1 + x)^{-3}$, his technique yields

$$1 + (-3)x + \frac{(-3)(-4)}{2}x^2 + \frac{(-3)(-4)(-5)}{6}x^3 + \cdots$$

or simply

$$(1 + x)^{-3} = 1 - 3x + 6x^2 - 10x^3 + 15x^4 - \cdots$$

where the series on the right never terminates. Using the definition of the negative exponents, this equation becomes

$$\frac{1}{(1+x)^3} = 1 - 3x + 6x^2 - 10x^3 + 15x^4 - \cdots \qquad \text{or equivalently}$$

$$\frac{1}{1 + 3x + 3x^2 + x^3} = 1 - 3x + 6x^2 - 10x^3 + 15x^4 - \cdots$$

This result Newton checked by cross-multiplying and canceling to confirm that indeed

$$(1 + 3x + 3x^2 + x^3)(1 - 3x + 6x^2 - 10x^3 + 15x^4 - \cdots) = 1$$

Things became even more peculiar when he expanded an expression like $\sqrt{1-x} = (1-x)^{1/2}$. In this case, $Q = -x$ and $m/n = 1/2$, so we get

$$\sqrt{1-x} = 1 + \frac{1}{2}(-x) + \frac{(1/2)(-1/2)}{2}(-x)^2 + \frac{(1/2)(-1/2)(-3/2)}{6}(-x)^3 + \cdots$$

$$= 1 - \frac{1}{2}x - \frac{1}{8}x^2 - \frac{1}{16}x^3 - \frac{5}{128}x^4 - \frac{7}{256}x^5 - \cdots \quad (*)$$

To check this peculiar-looking formula, Newton multiplied the infinite series on the right by itself—in short, he squared it—as follows:

$$\left(1 - \frac{1}{2}x - \frac{1}{8}x^2 - \frac{1}{16}x^3 - \frac{5}{128}x^4 - \cdots\right)\left(1 - \frac{1}{2}x - \frac{1}{8}x^2 - \frac{1}{16}x^3 - \frac{5}{128}x^4 - \cdots\right) = 1 - \frac{1}{2}x - \frac{1}{2}x - \frac{1}{8}x^2 + \frac{1}{4}x^2$$

$$- \frac{1}{8}x^2 - \frac{1}{16}x^3 + \frac{1}{16}x^3 + \frac{1}{16}x^3 - \frac{1}{16}x^3 - \cdots$$

$$= 1 - x + 0x^2 + 0x^3 + 0x^4 + \cdots = 1 - x$$

Hence

$$\left(1 - \frac{1}{2}x - \frac{1}{8}x^2 - \frac{1}{16}x^3 - \frac{5}{128}x^4 - \cdots\right)^2 = 1 - x$$

which verified that

$$1 - \frac{1}{2}x - \frac{1}{8}x^2 - \frac{1}{16}x^3 - \frac{5}{12}x^4 - \cdots = \sqrt{1-x}$$

as Newton had claimed.

"Extraction of roots are much shortened by this theorem," wrote Newton. That is, suppose we seek a decimal approximation to $\sqrt{7}$. First observe that

$$7 = 9\left(\frac{7}{9}\right) = 9\left(1 - \frac{2}{9}\right)$$

and so $$\sqrt{7} = \sqrt{9\left(1 - \frac{2}{9}\right)} = 3\sqrt{1 - \frac{2}{9}}$$

Now replace the square root by the first six terms in the binomial expansion labeled (*) above, where of course 2/9 is playing the role of x. We thereby get

$$\sqrt{7} \approx 3\left(1 - \frac{1}{9} - \frac{1}{162} - \frac{1}{1458} - \frac{5}{52488} - \frac{7}{472392}\right) = 2.64576\ldots$$

This result differs from the true value of $\sqrt{7}$ by only .00001, which is certainly quite impressive for using just six numerical terms. If we carried the binomial to more terms, we could guarantee greater accuracy of the estimation. Moreover, the same technique will provide approximate cube roots, fourth roots, and so on, since we could just as well apply the binomial theorem to expand $\sqrt[3]{1 - x} = (1 - x)^{1/3}$ and proceed as above.

In one sense, there is nothing terribly surprising about the fact that $\sqrt{7}$ can be approximated by a sum of six fractions. The truly amazing thing about this whole procedure is that Newton's binomial theorem shows us precisely *which* fractions to use, and generates them in an utterly mechanical fashion, devoid of the need for any particular insight or ingenuity on our part. It was a remarkably efficient and clever way to get roots of any order.

The binomial theorem is one of two prerequisites for the great theorem we shall soon examine. The other is Newton's inverse fluxions, or what we today call integration. A thorough explanation of inverse fluxions would carry us beyond the scope of this book and into the realm of calculus. However, we can state the key result in Newton's words and illustrate with an example or two.

It appeared in *De Analysi,* a treatise Newton composed in the middle of 1669 but did not publish until 1711. This was Newton's first extended treatment of his fluxional ideas, and he circulated it to a few mathematical colleagues. We know, for instance, that Isaac Barrow saw it, for he wrote to an acquaintance on July 20, 1669, that ". . . a friend of mine . . . that hath a very excellent genius to these things, brought me the other day some papers." The very first rule that Barrow or any other reader of *De Analysi* would encounter was the following.

> Let the Base AB of any Curve AD have BD for its perpendicular Ordinate; and call $AB = x$, $BD = y$, and let a, b, c, etc. be given Quantities, and m and n whole Numbers. Then:
>
> Rule 1: If $ax^{m/n} = y$, it shall be $\dfrac{an}{m+n} x^{(m+n)/n} =$ Area ABD.

O F

A N A L Y S I S

B Y

Equations of an infinite Number of Terms.

1. *THE General Method, which I had devised some considerable Time ago, for measuring the Quantity of Curves, by Means of Series, infinite in the Number of Terms, is rather shortly explained, than accurately demonstrated in what follows.*

2. Let the Base AB of any Curve AD have BD for it's perpendicular Ordinate; and call AB=x, BD=y, and let a, b, c, &c. be given Quantities, and m and n whole Numbers. Then

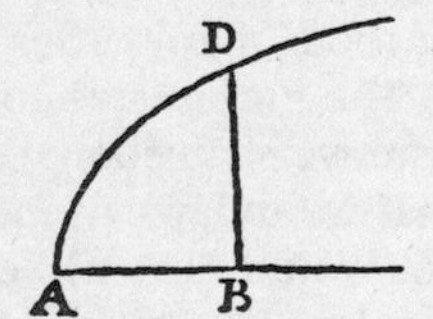

The Quadrature of Simple Curves,

RULE I.

3. If $ax^{\frac{m}{n}}=y$; it shall be $\frac{an}{m+n}x^{\frac{m+n}{n}}$ =Area ABD.

Newton's rule for finding areas under curves, from a 1745 translation of *De Analysi*
(photograph courtesy of Johnson Reprint Corporation)

In Figure 7.1, Newton was finding the area above the horizontal axis, beneath the curve $y = ax^{m/n}$, and as far to the right as the point x. According to Newton, this area was $\frac{an}{m+n}x^{(m+n)/n}$. For instance, if we take the straight line $y = x$ (Figure 7.2), where $a = m = n = 1$, this formula yielded an area $\frac{1}{2}x^2$, which is easily verified by the formula for

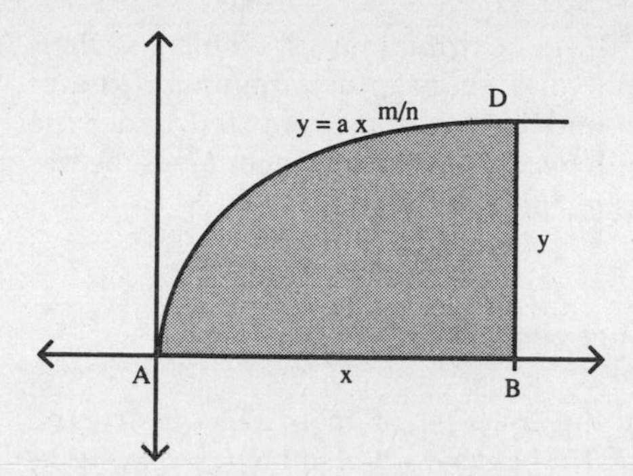

FIGURE 7.1

triangular area = ½(base) × (height). Similarly, the area under $y = x^2$ between the origin and the point x is $x^{2+1}/(2 + 1) = x^3/3$.

Further, as Newton explained in Rule 2 of *De Analysi*, "If the Value of y be made up of several such Terms, the Area likewise shall be made up of the Areas which result from every one of the Terms." As an example, he noted that the area beneath the curve $y = x^2 + x^{3/2}$ is just

$$\tfrac{1}{3}x^3 + \tfrac{2}{5}x^{5/2}$$

These, then, were to be Newton's tools: the binomial theorem and this fluxional method of finding areas under certain curves. These tools served him well in attacking any number of extraordinary mathematical and physical problems, but we shall watch Newton employ them to shed an entirely new light on an age-old problem: the estimation of the value of π. In the Epilogue to Chapter 4, we traced some of the history of this

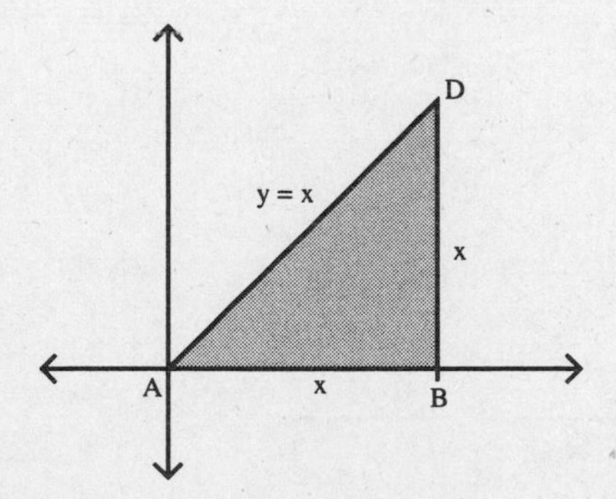

FIGURE 7.2

famous number and recognized the contributions of scholars such as Archimedes, Viète, and Ludolph von Ceulen in determining π to ever greater accuracy. Sometime around 1670, the problem fell under the gaze of Isaac Newton. Armed with his wonderful new techniques, he was about to launch a brilliant attack on an old foe.

Great Theorem: Newton's Approximation of π

Newton had certainly mastered the concepts of analytic geometry, and he cast his work in this format. He began with a semicircle having its center C at $(\frac{1}{2}, 0)$ and radius $r = \frac{1}{2}$, as shown in Figure 7.3. He knew that the circle's equation was

$$(x - \tfrac{1}{2})^2 + (y - 0)^2 = \tfrac{1}{2}^2 \qquad \text{or} \qquad x^2 - x + \tfrac{1}{4} + y^2 = \tfrac{1}{4}$$

Simplifying and solving for y gives the equation of the upper semicircle as

$$y = \sqrt{x - x^2} = \sqrt{x}\,\sqrt{1 - x} = x^{1/2}(1 - x)^{1/2}$$

(Exactly why he chose *this* particular semicircle may seem a complete mystery, but its special utility will in the end become clear.)

As shown earlier in equation (*), the expression $(1 - x)^{1/2}$ can be replaced by its binomial expansion, thus giving the equation of the semicircle as

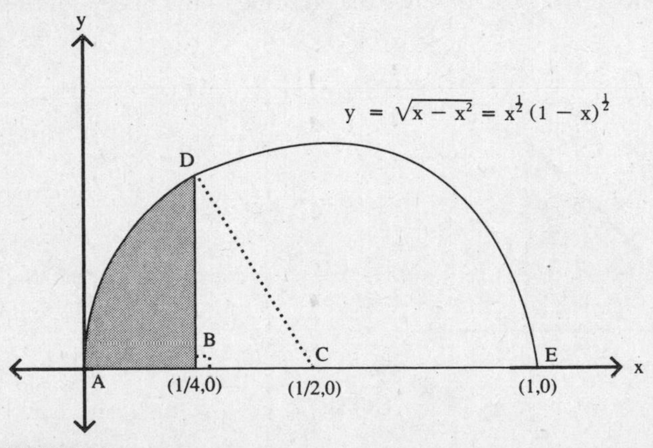

FIGURE 7.3

$$y = x^{1/2}(1-x)^{1/2}$$
$$= x^{1/2}\left(1 - \frac{1}{2}x - \frac{1}{8}x^2 - \frac{1}{16}x^3 - \frac{5}{128}x^4 - \frac{7}{256}x^5 - \cdots\right)$$
$$= x^{1/2} - \frac{1}{2}x^{3/2} - \frac{1}{8}x^{5/2} - \frac{1}{16}x^{7/2} - \frac{5}{128}x^{9/2} - \frac{7}{256}x^{11/2} - \cdots$$

And now, the genius of Isaac Newton becomes apparent. He let B be the point (¼, 0), as indicated in Figure 7.3, and drew BD perpendicular to the semicircle's diameter AE. He then attacked the shaded area ABD in two very different ways:

1. Area (ABD) by fluxions As we have seen, Newton knew how to find the area under such a curve from its starting point at 0 rightward to the point $x = \frac{1}{4}$. That is, by Rules 1 and 2 of *De Analysi*, the shaded area was just

$$\frac{2}{3}x^{3/2} - \frac{1}{2}\left(\frac{2}{5}x^{5/2}\right) - \frac{1}{8}\left(\frac{2}{7}x^{7/2}\right) - \frac{1}{16}\left(\frac{2}{9}x^{9/2}\right) - \cdots$$
$$= \frac{2}{3}x^{3/2} - \frac{1}{5}x^{5/2} - \frac{1}{28}x^{7/2} - \frac{1}{72}x^{9/2} - \frac{5}{704}x^{11/2} - \cdots \qquad (**)$$

evaluated for the value $x = \frac{1}{4}$. The genius of his approach is that the resulting expression simplifies beautifully when we evaluate it, since

$$\left(\frac{1}{4}\right)^{3/2} = \left(\sqrt{\frac{1}{4}}\right)^3 = \frac{1}{8}, \quad \left(\frac{1}{4}\right)^{5/2} = \left(\sqrt{\frac{1}{4}}\right)^5 = \frac{1}{32}, \quad \text{and so on}$$

Thus we approximate the shaded Area (ABD), using the first *nine* terms of series (**), by

$$\frac{1}{12} - \frac{1}{160} - \frac{1}{3584} - \frac{1}{36864} - \frac{5}{1441792} \cdots - \frac{429}{163208757248} = .07677310678$$

2. Area (ABD) by geometry Newton next reexamined the problem of the shaded area from a purely geometric perspective. He first determined the area of right triangle $\triangle DBC$. Notice that the length of BC is ¼, while CD, being a radius, has length $r = \frac{1}{2}$. A direct application of the Pythagorean theorem yielded

$$\overline{BD} = \sqrt{\left(\frac{1}{2}\right)^2 - \left(\frac{1}{4}\right)^2} = \sqrt{\frac{3}{16}} = \frac{\sqrt{3}}{4}$$

Hence,

$$\text{Area }(\triangle DBC) = \frac{1}{2}(\overline{BC}) \times (\overline{BD}) = \frac{1}{2}\left(\frac{1}{4}\right)\left(\frac{\sqrt{3}}{4}\right) = \frac{\sqrt{3}}{32}.$$

So far, so good. Next Newton wanted the area of the wedge- or pie-shaped sector *ACD*. To determine this area, he again referred to $\triangle DBC$. With the length of *BC* being exactly half that of the hypotenuse *CD*, he recognized this as a familiar 30°-60°-90° right triangle; in particular, $\angle BCD$ was a 60° angle.

Again, one is struck by his penetrating insight, for by placing his perpendicular at a point other than *B*, he would not have emerged with a simple 60° angle when he needed it most. But, knowing that the angle of the sector was 60°—that is, one-third of the 180° angle forming the semicircle—Newton could see that the area of the sector was likewise a third of the area of the semicircle. In short,

$$\text{Area (sector)} = \frac{1}{3}\text{ Area (semicircle)}$$

$$= \frac{1}{3}\left(\frac{1}{2}\pi r^2\right) = \frac{1}{3}\left[\frac{1}{2}\pi\left(\frac{1}{2}\right)^2\right] = \frac{\pi}{24}$$

The perceptive reader, recalling that this great theorem was to have been Newton's approximation of π, may have been worrying about how and when this constant was ever going to enter the argument. At last, π has appeared in Newton's chain of reasoning, and there now remains just a final step or two to get a wonderfully efficient approximation of it.

Thus, the geometric approach to the shaded area yields

$$\text{Area }(ABD) = \text{Area (sector)} - \text{Area }(\triangle DBC) = \frac{\pi}{24} - \frac{\sqrt{3}}{32}$$

Equating this result with the fluxion/binomial theorem approximation for the same shaded area from above, we have $.07677310678 \approx \pi/24 - \sqrt{3}/32$ and solving for π gives us the estimate

$$\pi \approx 24\left(.07677310678 + \frac{\sqrt{3}}{32}\right) = 3.141592668\ldots$$

Q.E.D.

The amazing thing about this estimate is that, with just *nine* terms of the binomial expansion, we have found π correct to seven decimal places, and our estimate above differs from the true value of π by less than .000000014. This represents a major advance over the awesome calculations of Viète or Ludolph mentioned in Chapter 4. In fact, the only real difficulty with this technique is the need for an accurate estimate of $\sqrt{3}$. But, as we have seen previously, Newton's binomial theorem handled the computation of square roots with ease. In short, this result clearly demonstrated the efficacy of his new mathematical discoveries in addressing an old problem with remarkable success.

Newton's approximation of π is taken directly from his *Methodus Fluxionum et Serierum Infinitarum,* a treatise written in 1671 but not published for decades, that extended the fluxional ideas of *De Analysi* from a few years earlier. In *Methodus*, Newton presented the value of π to 16 decimal places, based on the 20-term binomial expansion of $\sqrt{1 - x}$. At one point, commenting on such approximations, he somewhat sheepishly confessed, "I am ashamed to tell you to how many places of figures I carried these computations, having no other business at the time."

Newton's shame notwithstanding, those who can appreciate the subtle beauty of mathematics will be grateful that no pressing business otherwise occupied his fertile mind while he ". . . was in my prime age for invention and minded Mathematicks and Philosophy more then at any time since."

Epilogue

These, then, were some of the fruits of Newton's student days at Trinity College during the plague years of the mid-1660s. But Isaac had a life of three-score years yet to live, and it would carry the unhappy child of a small English village to astonishing heights of fame and influence. The end of this chapter charts the remainder of his remarkable odyssey.

In 1668, Newton completed his master's degree and was elected a fellow of Trinity College. This meant that he could stay indefinitely in his academic post—with financial support—provided he took holy vows and remained unmarried. As if that were not prestigious enough, the following year Isaac Barrow resigned the Lucasian chair of mathematics and successfully urged the appointment of Newton as his successor. There is a charming legend to the effect that Barrow abdicated the Lucasian chair because he recognized Newton as his mathematical superior and thus could not retain the chair in good conscience. Actually, Barrow's motives were a bit less magnanimous, for he was also a brilliant scholar of Greek and theology and was angling for a higher position in a somewhat different line of work. In fact, having left the Lucasian chair, Barrow soon landed the position of Chaplain to the King. Nonetheless, he was instrumental in getting the appointment for his young and rather unknown colleague. Barrow certainly knew excellence when he encountered it and heartily recommended Newton as ". . . a fellow of our College, . . . very young . . . but of an extraordinary genius and proficiency."

Newton's duties as Lucasian professor were minimal. He was obliged neither to accept students nor to do any tutoring. The main job, besides picking up the substantial paycheck and remaining morally chaste, was to deliver regular lectures on mathematical topics. Those who imagine throngs of eager students flooding into the lecture hall to hear this great man will be in for a surprise. Recall that Newton was not yet a name to be reckoned with outside of a very small circle, and that Cambridge students were not necessarily disposed to the intense intellectual life. A contemporary recorded this account of Newton's Lucasian lectures:

> . . . so few went to hear Him, & fewer yet understood him, that oftimes he did in a manner, for want of Hearers, read to ye Walls.

The commentator added that Newton's lectures would last for half an hour except when there was no one at all in the audience; in that case he would stay only 15 minutes.

If Newton was unsuccessful as a lecturer, his scientific output was prodigious. He made few friends, keeping to himself as an aloof and somewhat strange character of Trinity College. An associate of many years recalled that there was but once that he saw Newton laugh. His mirth was prompted by an acquaintance who, examining a copy of Euclid, asked what conceivable value this decrepit old book might have. At this Newton roared with glee.

One of the best pictures of Newton as professor came from his nephew Humphrey Newton, who wrote:

> He always kept close to his studyes, very rarely went a visiting, & had as few Visiters. . . . I never knew him take any Recreation or Pastime, either in Riding out to take ye Air, Walking, Bowling, or any other Exercise whatever, Thinking all Hours lost, that was not spent in his studyes. . . . He very rarely went to Dine in ye Hall . . . & then, if He has not been minded, would go very carelessly, with Shooes down at Heels, Stockins unty'd, surplice on, & his Head scarcely comb'd.

Nonetheless, his reputation began to grow, mainly through the circulation of such unpublished treatises as *De Analysi* or *Methodus*. His first big public splash came in 1671 when he displayed his newly invented reflecting telescope at a meeting of the Royal Society in London. This optical device was the perfect vehicle for combining Newton's theories of light and his practical ability at tinkering. The scientific community applauded his efforts, and to this day reflecting telescopes, which rely on a mirror at the base rather than a heavy and unstable lens at the top, are the preferred instruments of optical astronomy.

Flushed with the success of this invention, Newton soon submitted a paper on optics to the Royal Society. This time, however, his radical ideas were met with skepticism and derision from such noted scholars as Robert Hooke. Controversy, which is a common feature of the scholar's world, was anathema to Newton. Stung by the criticism, he withdrew deeper into his private world, refusing to publish or communicate his ideas lest they ensnare him in further bickering with his less enlightened peers. This decision meant that brilliant scientific papers would lie unknown and unpublished in his desk drawer for decades. As the next chapter shows, this practice had disastrous consequences when, in subsequent years, he would claim priority for ideas—particularly the calculus—published first by others.

As the 1670s progressed, Newton's interests turned away from mathematics and physics. He devoted a great deal of time to alchemy, although we can see in his investigations the mind of a modern-day chemist at work. Less modern, perhaps, was his study of the Scriptures with an eye toward calculating the ages and dates of the prophets, the size of the Ark of the Covenant, and other such things. He spent untold hours meticulously analyzing the Bible in this fashion. One result of these researches may have been his rejection of the notion of Jesus as one of the Blessed Trinity—an odd twist, given the name of the College at which he was employed. His views were radical enough to force Isaac to keep silent, at least as long as he held the Lucasian chair.

This brings us to the year 1684. It was then that Edmund Halley, who later became famous for the comet that bears his name, visited Newton and urged him to publish some of his awesome discoveries. As always, Newton was reluctant, but Halley's prodding, not to mention his bearing the costs of publication, persuaded Newton that the time was right. He caught fire, working diligently on what would be his scientific masterpiece, a description of his researches into the laws of motion and the principle of universal gravitation. The work came out in 1687 with the title *Philosophiae Naturalis Principia Mathematica.* Here was the system of the universe, a precise mathematical derivation of the motions of the moon and planets, that explained the clockwork precision of all creation, ticking to the orderly rhythms of Newton's wonderful equations. After *Principia*, science would never be the same.

Principia was an enormous success. Although few could understand all the intricacies of the text, people universally came to regard Newton as being an almost superhuman phenomenon. Many years later, the French mathematician Pierre-Simon Laplace would record the respect, the awe, and the envy he felt toward Newton's scientific discoveries:

> Newton was the greatest genius that ever existed, and the most fortunate, for we cannot find more than once a system of the world to establish.

The year following publication of the *Principia* was a pivotal one in British history. Late in 1688, James II, the last of the Stuart kings, fled the throne, to be replaced by William III and Mary II. The subsequent political reform, called the "Glorious Revolution," saw the influence of Parliament rise and that of the monarch simultaneously fall. Interestingly, the Member of Parliament sent in 1689 from Cambridge to Westminster was none other than the Lucasian professor, Isaac Newton.

As a supporter of the new monarchs, Newton apparently made little impression on British government from his parliamentary seat. Nonetheless, his life had surely taken a new direction. No longer the solitary, isolated scholar, he had now emerged into the public arena in a manner inconceivable even a few years before. With the solid triumph of the *Principia* behind him, the Cambridge professor was becoming the London official. He seemed to enjoy the change, making friends with such notables as John Locke and Samuel Pepys. Although in 1693 he suffered a brief nervous breakdown— sometimes attributed to his common practice of tasting the chemical compounds he used in his alchemy experiments—Newton was back on even keel by 1695 and a year later resigned the Lucasian chair and left Trinity College. It had been 35 years since he had arrived there as an ordinary undergraduate from Woolsthorpe, and those three-and-a-half decades had transformed this young man in a way that no one could have anticipated.

And what did the ex-professor do then? With a favorable impression of public service and perhaps a growing realization that his days of scientific productivity were waning, Newton was ready to try something radically different. He thus accepted the position of Warden of the Mint in 1696. At that time, British coins were produced at the Tower of London, where the warden both lived and worked. By all accounts, Newton did a fine job at the Mint, overseeing a general recoinage of English currency and mixing well with the financiers and bankers of the City of London.

These years also gave Newton the opportunity to attend to his scientific writings. In 1704, he published his *Opticks*, a massive tome that did for his optical theories what the *Principia* had done for his law of gravitation. Interestingly, it was as an appendix to the *Opticks* that New-

Isaac Newton as Lucasian Professor at Cambridge (photograph courtesy of Yerkes Observatory, University of Chicago)

ton first published an account of his fluxional methods, a work called *De Quadratura*. Although he had developed these ideas four decades earlier, it was only in 1704 that the world saw them in print. Unfortunately, this was a bit too late. Mathematicians on the continent had published their own research on the calculus years before. Some continental mathematicians reacted with indifference if not outright skepticism to Newton's claim that the ideas he was now publishing were really 40 years old.

Just prior to the *Opticks*, Newton had been elected president of the Royal Society. He brought to this position the same impressive administrative skills that had become so evident at the Mint. Newton retained the position at the head of the Society until his death.

Unmatched scientist, pre-eminent mathematician, public servant, and Royal Society president, Isaac Newton received the ultimate tribute of a knighthood from Queen Anne in 1705. Fittingly, the ceremony occurred at Trinity College, Cambridge. He still had 22 years to live with the honored title of "Sir Isaac."

He lived these years in London, dividing his time among duties at the Mint and the Royal Society, his scientific writings, and the trappings of influence in the British capital. These must have been heady years for Sir Isaac, and his power and reputation—not to mention his personal fortune—grew constantly.

He lived to the age of 84, and at his death in 1727, Isaac Newton was regarded as a national treasure, as indeed he was. Clearly the reigning scientist in Europe, his impact had been nothing short of revolutionary. In death he was accorded the signal honor of burial, along with kings and military heroes, in Westminster Abbey. Today, Newton's statue stands prominently, at the left-hand portal of the Abbey's great Choir Screen, visible to all who enter that revered place.

The world was abundant in its praise of this man. For instance, Alexander Pope, the great British poet, was moved to write:

> Nature and Nature's Laws lay hid in Night.
> God said, "Let Newton be" and all was light.

Another famous poet, William Wordsworth, was a bit more restrained when he wrote of a night spent at Trinity College:

> And from my pillow, looking forth by light
> Of moon or favouring stars, I could behold
> The antechapel where the statue stood
> Of Newton with his prism and silent face,
> The marble index of a mind for ever
> Voyaging through strange seas of Thought, alone.

It would be difficult to overestimate the influence of this solitary voyager. One need only recall the world-view of Cardano from a century before—a peculiar blend of hard science and the most outlandish superstition. At that time, the world was seen largely as an irrational place, with supernatural agencies intervening in everything from the appearance of comets to the daily calamities of life. Newton, with his clockwork universe, removed the supernatural from Nature. His work described a rational world, one obeying fundamental laws which—and this was no small part of Newton's legacy—could be deciphered by mere mortals.

It is interesting to observe that, 166 years after Newton's arrival at Cambridge, another British youth began his undergraduate career at the University's Christ College, just a few blocks from Newton's old rooms at Trinity. The young Charles Darwin undoubtedly walked the same Cambridge streets that Newton had known so many years before. Like

Newton's portrait from the British 1-pound note

Newton, Darwin was also reluctant to publish his discoveries, but when he put pen to paper, he wrote the 1859 classic *Origin of Species*, a masterpiece that did for biology what the *Principia* had done for physics. Even as Newton had made the physical world "natural," so did Darwin make the biological world "natural" in explaining the mechanism for the seemingly unexplainable dynamism of life on this planet. Both men influenced far, far more than mere science, and both saw their theories affect the human perception of reality in a profound and revolutionary fashion. Today Darwin also lies in Westminster Abbey, a few feet from the Newton monument—two scientific giants, two Cambridge men who reached the top.

Late in his life, Isaac Newton looked back upon his remarkable intellectual adventure and graciously acknowledged that, if he had seen farther than others, it was because he had stood on the shoulders of giants. He was, of course, thinking of Viète, Galileo, Descartes, and other figures of the heroic century. Now his own shoulders would be available for the scholars of future generations. In an often quoted, and quite striking reflection, Newton mused:

> I do not know what I may appear to the world; but to myself I seem to have been only like a boy playing on the seashore, and diverting myself in now and then finding a smoother pebble or a prettier shell than ordinary, whilst the great ocean of truth lay all undiscovered before me.

But perhaps we should leave him, finally at rest in Westminster Abbey, with this fitting epitaph:

> Mortals, congratulate yourselves that so great a man has lived for the honor of the human race.

8

Chapter

The Bernoullis and the Harmonic Series

(1689)

The Contributions of Leibniz

While the solitary Isaac Newton was changing the face of mathematics from his Cambridge rooms, mathematicians on the European Continent were far from idle. Influenced by the work of Descartes, Pascal, and Fermat, continental mathematics flourished during the latter half of the seventeenth century. Its greatest practitioner by far was Gottfried Wilhelm Leibniz (1646–1716).

Often described as a universal genius, Leibniz mastered a bewildering array of scholarly subjects and left his mark upon each. He began as a child prodigy with access to the sizable library of his father, a professor of moral philosophy. Young Gottfried made the most of this opportunity, teaching himself Latin and Greek at an early age and consuming the books on his father's shelves so voraciously that he was ready to enter the University of Leipzig at the age of 15. He blazed through his university experience rapidly and, when barely 20, had completed his doctoral dissertation at Altdorf.

Although on the brink of a promising academic career, Leibniz left the university to work for the Elector of Mainz, a potentate of one of the many small states into which Germany was then divided. On the job he examined certain complex legal matters, including a major reform of the Holy Roman Empire, while in his spare time he designed a calculating machine that was to multiply by repeated, rapid additions and divide by similarly rapid subtractions. Although the technology of the day limited the performance of such a device—an embarrassment to Leibniz, who had touted his machine's efficiency—the theory was solid and, eventually, workable.

In 1672, Leibniz was sent from Germany to Paris as a high-level diplomat. He was intoxicated by the intellectual life of the French capital, and side trips to London and Holland brought the young genius into contact with such established scholars as Hooke, Boyle, van Leeuwenhoek, and the philosopher Spinoza. Leibniz certainly found himself in a lively academic environment. Yet in 1672, even he would have admitted that his mathematical training was limited to the masterpieces of classical times. Leibniz, with such talent and curiosity, needed a "crash course" on the current trends and directions of mathematics.

Fortunately, he found the perfect opportunity in Paris. It came in the person of the Dutch scientist Christiaan Huygens (1629–1695), who had been living in Paris under a pension from the Sun King, Louis XIV. Huygens' credentials were impressive. On the theoretical side, he had done extensive work on mathematical curves, especially one known as the "cycloid." This is the path traced by a point fixed to the rim of a circle that is rolling along a horizontal path (see Figure 8.1). His discoveries played a role in his design of the first successful pendulum clock, whose internal operations were intimately linked to cycloidal curves.

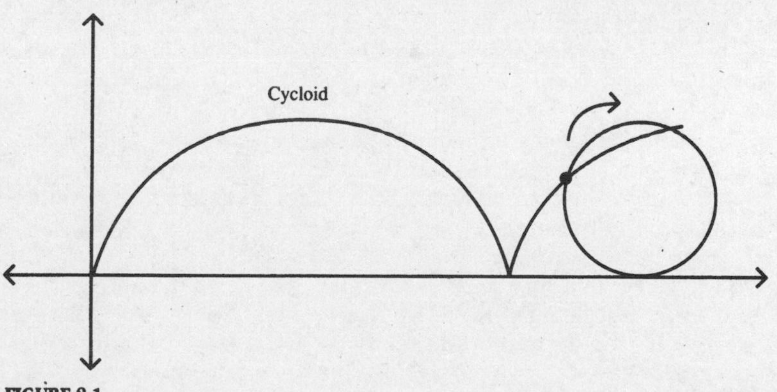

FIGURE 8.1

As this invention suggests, Huygens was interested in areas other than pure mathematics. Indeed it is perhaps in physics and astronomy that he built his most enduring reputation, with his investigation of the laws of motion, his study of centrifugal forces, and his proposal of a sophisticated wave theory of light. Moreover, Huygens was the first to explain that the bizarre appendages appearing in telescopic views of the planet Saturn were actually rings.

With such a scientific resource available on his Parisian doorstep, it is little wonder that Leibniz sought out Huygens for advice on strengthening his own mathematical training. It probably overstates the matter to suggest that Huygens was Leibniz's teacher, but he did guide and direct the young diplomat in the study of current mathematical problems, and certainly few teachers in history have had a student like Gottfried Wilhelm Leibniz.

One problem Huygens suggested to Leibniz was the determination of the sum of the reciprocals of the so-called triangular numbers. These are numbers that correspond to triangular arrays of objects, as shown in Figure 8.2. The first such number is 1, the second is 3, the third is 6, and in general the kth triangular number has the form $k(k + 1)/2$. Note that, in the game of bowling, the desire for a wedge-shaped array of pins at the end of the alley translates into a set of ten of them, an obvious "triangular number."

Huygens wanted Leibniz to evaluate the sum, not of the triangular numbers, but of their reciprocals. In short, he was challenging his young protégé to determine the value S, where

$$S = 1 + 1/3 + 1/6 + 1/10 + 1/15 + 1/21 + \cdots$$

After some thought, Leibniz divided all terms by 2 to get

$$1/2 S = 1/2 + 1/6 + 1/12 + 1/20 + 1/30 + \cdots$$

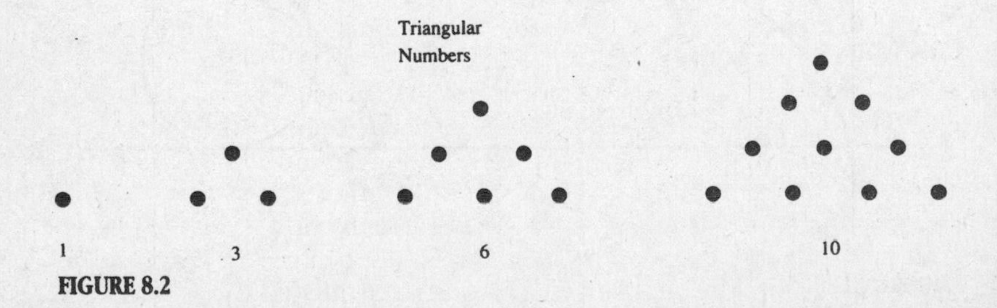

FIGURE 8.2

in which he noticed a striking pattern. For, he could replace the $\frac{1}{2}$ that began the right-hand side by the equivalent $1 - \frac{1}{2}$; next he could replace the $\frac{1}{6}$ by $\frac{1}{2} - \frac{1}{3}$; the $\frac{1}{12}$ by $\frac{1}{3} - \frac{1}{4}$; and so on. This transformed the expression above into

$$\tfrac{1}{2}S = (1 - \tfrac{1}{2}) + (\tfrac{1}{2} - \tfrac{1}{3}) + (\tfrac{1}{3} - \tfrac{1}{4}) + (\tfrac{1}{4} - \tfrac{1}{5}) + \cdots$$

Leibniz then simply removed the parentheses and canceled with abandon to get

$$\tfrac{1}{2}S = 1 - \tfrac{1}{2} + \tfrac{1}{2} - \tfrac{1}{3} + \tfrac{1}{3} - \tfrac{1}{4} + \tfrac{1}{4} - \tfrac{1}{5} + \cdots = 1$$

But, if half of S is 1, then S itself (the sum of the reciprocals of the triangular numbers) is evidently 2. In short, Leibniz had cleverly solved Huygens' challenge and found that

$$1 + \tfrac{1}{3} + \tfrac{1}{6} + \tfrac{1}{10} + \tfrac{1}{15} + \tfrac{1}{21} + \cdots = 2$$

Modern mathematicians voice certain reservations about his cavalier manipulations of infinite series in this argument, but no one can deny the basic ingenuity of his approach.

This was just the beginning of Leibniz's mathematical insights. He soon applied his great talents to the same questions about tangents and areas that Newton had addressed a decade earlier. By the time Leibniz had left Paris in 1676, he had discovered for himself the fundamental principles of calculus. His four Parisian years had seen him rise from mathematical novice to mathematical giant.

But even as these years lay the foundation for his enduring fame, they likewise lay the foundation for an enduring controversy. We recall that Isaac Newton's fluxions were known only to a select number of English mathematicians who had seen his hand-written manuscripts on the subject. On a visit to London, to be inducted into Britain's Royal Society in 1673, Leibniz saw some of these Newtonian documents and was greatly impressed. Later, in correspondence that was routed through Henry Oldenburg, the Royal Society secretary, Leibniz inquired further about Newton's discoveries, and the great Englishman answered, albeit in a somewhat veiled fashion, in two famous 1676 letters, now called the *epistola prior* and the *epistola posterior*. These Leibniz read intently.

And so it was that, when Gottfried Wilhelm Leibniz published his first paper on this amazing new mathematical method, his British counterparts cried "foul!" The paper carried the lengthy title "Novo Methodus pro Maximis et Minimis, itemque tangentibus, qua nec fractas, nec irrationales quantitates moratur, et singulare pro illis calculi genus" ("A

New Method for Maxima and Minima, as well as Tangents, which is impeded neither by Fractional nor Irrational Quantities, and a Remarkable Type of Calculus for this"). It appeared in 1684 in the scholarly journal *Acta Eruditorum,* of which Leibniz happened to be editor.

Thus the world learned the calculus from Leibniz and not from Newton. Indeed, it was from the title of this paper that the subject took its name. But the British, champions of their countryman, asserted with varying degrees of tact that Leibniz had plagiarized the whole business. His visit to England, his familiarity with the quietly circulating Newtonian manuscript, his exchange of letters—all convinced British partisans that the scoundrel Leibniz had stolen Newton's glory.

The bickering that followed does not constitute one of the more admirable chapters in the history of mathematics. At first, the two principals tried to stay above the fray, with subordinates battling on their behalf. Eventually, however, all parties ended up becoming involved, usually to no good end. Leibniz freely admitted his contact with Newtonian ideas through letters and manuscripts, but observed that these had given him only hints at results, not clear-cut methods; these methods Leibniz discovered for himself.

Meanwhile, the English became increasingly furious. Worse (from the British viewpoint), Leibniz's calculus caught on quickly in Europe and was amplified by some of his disciples, whereas the isolated Newton still refused to publish anything on the subject. Recall that Newton had written his first tract on fluxions in October 1666, almost two decades before Leibniz's paper rolled off the presses; yet it was not until 1704 that Newton published a specific account of his method in the appendix to his *Opticks. De Analysi,* a more thorough account of the subject that had been informally circulating in England's mathematical community at the time of Leibniz's 1673 visit, did not appear in print until 1711. A full-blown development of Newton's ideas, carefully and fully written by the author in order to be "a compleat institution for the use of learners," appeared only in 1736, a full nine years after Sir Isaac's death! In fact, so tardy was Newton in publishing his mathematics that some of Leibniz's more ardent supporters could claim that it was *Newton* who did the plagiarizing from the already published works of Leibniz, and not vice versa.

Clearly, the situation was a mess. In his book *Philosophers at War,* Rupert Hall gives a fascinating and detailed account of the charges and countercharges that flew back and forth across the Channel. Today, now that the smoke has had almost three centuries to clear, it is recognized that both men deserve credit for independently developing virtually the same body of ideas. The simultaneous discovery of important concepts is not at all uncommon in science, as mentioned in the discussion of the

Gottfried Wilhelm Leibniz (photograph courtesy of The Ohio State University Libraries)

origins of non-Euclidean geometry in Chapter 2. A century and a half after the Newton/Leibniz controversy, the biological world saw the simultaneous appearance of the theory of natural selection from Englishmen Alfred Russel Wallace and Charles Darwin. In this latter instance, Darwin's enormously influential *Origin of Species* contrasted with Wallace's less ambitious writings and may account for Darwin's enduring fame. Further, the fact that the co-discoverers of evolution were both British eliminated the unfortunate nationalistic overtones of the Newton/Leibniz flare-up.

When not engaged in the controversy over the origins of the calculus, Leibniz devoted his time to the remarkable variety of pursuits that had characterized his life. He accepted a position with the Duke of Brunswick and attempted to trace that nobleman's distant genealogy. He became an expert in the Sanskrit language and the culture of China. Further, he continued to work in philosophy, a discipline that was always near to his heart. It was Leibniz who sought to develop a perfect system of formal logic, based on an "alphabet of human thought" and governed by a carefully prescribed "rational calculus." With such logical tools, Leibniz hoped that mankind could rid everyday life of its pervasive imprecision and irrationality. Of course, he never came close to succeeding in what can only be called a grandiose plan, but his attempts constituted the first real steps toward what we today call "symbolic logic." In particular, his use of algebraic formulas to denote logical statements was a significant advance beyond the verbal syllogisms of Greek logical theory.

In 1700, Leibniz was a major force in the creation of the Berlin Academy. This community of scholars, writers, and musicians was meant to bring to Berlin the greatest thinkers in Europe and thereby put that city on the intellectual map. Leibniz was honored by being named the Academy's president, a position he held for the remainder of his life.

In spite of the demands of the Berlin Academy, his work proceeded. He continued his studies in logic and philosophy, while simultaneously advocating reforms in the world's religious and political structures that he hoped would bring true peace and harmony among men. Ironically, his patron during these last years was a Hanoverian nobleman who, upon the death of England's Queen Anne in 1714, was tapped to become Britain's King George I. Leibniz longed to accompany King George to England and assume a position as official court historian, but George never extended the offer. It would certainly have proved fascinating for the two protagonists in the calculus battle—Newton and Leibniz—simultaneously to have been living in London. Unfortunately, it was not to be.

Leibniz died in 1716. Many friends and colleagues from the Hanoverian Court had gone to England; his own status had diminished somewhat; and reports are that only a trusted servant attended the funeral of this great man. This stands in stark contrast to Newton's titanic reputation in England, a reputation that, as noted in the previous chapter, led to his burial in the honored shrine of Westminster Abbey. While the apotheosis of Newton was certainly justified, Leibniz deserved a similar glory of his own.

In any comparison of the two great inventors of the calculus, one fact stands out. Newton, in a certain sense, took his fluxions with him to the grave. To his last day, the solitary, misanthropic Sir Isaac never surrounded himself with a crowd of talented disciples, eager to learn, refine, and extend his work. By contrast, it was Leibniz's good fortune that two of his most enthusiastic followers were Jakob and Johann Bernoulli of Switzerland, brothers who would become the major force behind disseminating and promoting the calculus throughout Europe. Their efforts, perhaps as much as those of Leibniz himself, gave the subject the flavor and appearance that it retains to this day.

The Brothers Bernoulli

Jakob Bernoulli (1654–1705), the elder of the two brothers, was a gifted mathematician who made important contributions to the calculus, to the summation of infinite series, and perhaps most notably to the emerging subject of probability. We have noted how this mathematical subdiscipline got its start in the sixteenth century with the work of Cardano and how it grew in sophistication with the Fermat-Pascal correspondence in the mid-seventeenth century. The posthumous publication of Jakob's *Ars Conjectandi* in 1713 ranks as the next milestone of probability theory. This massive work consolidated earlier discoveries and pushed the frontiers of the subject to new levels. It remains Jakob Bernoulli's masterpiece.

Meanwhile, younger brother Johann (1667–1748) was building his own mathematical reputation. With undisguised zeal, Johann Bernoulli took it upon himself to spread Leibniz's calculus across the continent. Johann, frequently in correspondence with his German mentor, was ever ready to defend Leibniz's reputation in the controversy with the Newtonian British. Recalling that Thomas Huxley earned the epithet "Darwin's Bulldog" for stoutly defending the great naturalist from attacks of the religious community of the mid-nineteenth century, we might likewise call Johann Bernoulli "Leibniz's Bulldog" for much the same reason. Like Huxley, Johann sometimes supported his client with an almost shocking intensity; also like Huxley, Johann ultimately succeeded in his mission.

One of Johann's most important contributions came through his connections with the Marquis de l'Hospital (1661–1704). The latter was a French nobleman and amateur mathematician who very much wanted to learn this revolutionary new calculus. The marquis thus employed Johann Bernoulli to supply him with tracts on various aspects of the subject, as well as to provide him with any new mathematical discoveries of note. In a sense, it appears that l'Hospital bought the rights to Bernoulli's mathematical research. In 1696, l'Hospital collected the Bernoullian writings and published the first calculus text under the title *Analyse des infiniment petits (Analysis of the Infinitely Small)*. Written in the vernacular rather than in Latin, the book was almost exclusively Bernoulli's in all but the name on the title page.

Down through history, there have been many illustrious brother teams. From Agamemnon and Menelaus of the Trojan War up to Wilbur and Orville Wright of aviation fame, the past is full of brothers working together to noble ends. Certainly, Jakob and Johann constitute mathematics' most important fraternal success story, but it must be noted that their relationship was far from harmonious. Quite the contrary, each became the other's fiercest competitor in mathematical matters, until their attempts at one-upsmanship seem, in retrospect, almost comical. There are times when the Bernoullis are more reminiscent of the Marx brothers than of history's other famous siblings.

Consider, for instance, the question of the catenary curve. This is the shape assumed by a chain, fixed at two points, and hanging under its own weight. In a 1690 paper, Jakob, the elder brother with the long-established reputation, posed the problem of determining the nature—that is, the equation—of such a curve. This problem had been in existence for a great many years—Galileo had surmised that the curve was a parabola—but the matter remained open. Jakob felt that the new, wonderful techniques of calculus might provide the key to its solution.

Unfortunately, he got nowhere on this vexing problem. After a year's unsuccessful effort, Jakob was chagrined to see the correct solution published by his young brother Johann. For his part, the upstart Johann could hardly be considered a gracious winner, as seen in his subsequent recollection of the incident:

> The efforts of my brother were without success; for my part, I was more fortunate, for I found the skill (I say it without boasting, why should I conceal the truth?) to solve it in full. . . . It is true that it cost me study that robbed me of rest for an entire night . . . but the next morning, filled with joy, I ran to my brother, who was still struggling miserably with this Gordian knot without getting anywhere, always thinking, like Galileo, that the catenary was a parabola. Stop! Stop! I say to him, don't torture yourself any more to try to prove the identity of the catenary with the parabola, since it is entirely false.

It is amusing to note the time required for Johann's successful solution. To sacrifice the rest "of an entire night" on a problem that Jakob had wrestled with for a year qualifies as a first-class insult if ever there was one.

In this chapter, we shall examine a great theorem in which both Bernoullis (perhaps in the midst of a rare truce) had a hand. The issue in question was the nature of the so-called "harmonic series," an infinite series with a very peculiar property. While we have already seen Leibniz's attack on a particular series, we should first make a few remarks about the question of infinite series in general.

In the seventeenth century, an infinite series was viewed merely as the sum of an endless collection of terms. Of course, there was no guarantee that such a series would have a *finite* sum; if we consider an example like $1 + 2 + 3 + 4 + 5 + \cdots$, it is clear that, as we proceed along, the sum will grow in size beyond any finite quantity. We say that such a series "diverges to infinity."

Johann Bernoulli (photograph Courtesy of Georg Olms Verlagsbuchhandlung)

On the other hand, there are series of infinitely many terms that sum to a finite number, a phenomenon that may at first appear paradoxical but which, upon reflection, seems reasonable enough. For instance, when we write the familiar decimal expansion ⅓ = .3333333 . . . , we mean precisely

$$\frac{1}{3} = \frac{3}{10} + \frac{3}{100} + \frac{3}{1000} + \frac{3}{10000} + \cdots$$

Leibniz's series cited above exhibits similar behavior, for the infinitely many terms in question sum to the (finite) number 2. We say such a series "converges," which means, in an informal sense, that its sum zeroes in on a particular number as we add ever more terms together.

Undoubtedly, the most important convergent series in mathematics is the geometric series. This is a series of the form

$$\alpha + \alpha^2 + \alpha^3 + \alpha^4 + \cdots + \alpha^k + \cdots$$

where we insist that $-1 < \alpha < 1$. Thus, a geometric series is the sum of α and all of its higher powers. A "seventeenth-century style" argument to verify its convergence runs as follows:

Let $S = \alpha + \alpha^2 + \alpha^3 + \alpha^4 + \cdots$ be the sum we are seeking. Multiplying both sides of this equation by α, we see that $\alpha S = \alpha^2 + \alpha^3 + \alpha^4 + \alpha^5 + \cdots$, and subtracting one from the other yields

$$S - \alpha S = (\alpha + \alpha^2 + \alpha^3 + \alpha^4 + \cdots) - (\alpha^2 + \alpha^3 + \alpha^4 + \alpha^5 + \cdots) = \alpha$$

since all terms cancel but the first. Consequently, $S(1 - \alpha) = \alpha$, and so $S = \alpha/(1 - \alpha)$. But S was just the sum of the original geometric series. Hence we can conclude:

$$\alpha + \alpha^2 + \alpha^3 + \alpha^4 + \alpha^5 \cdots = \frac{\alpha}{1 - \alpha}$$

For instance, if $\alpha = 1/3$, we see that

$$\frac{1}{3} + \frac{1}{9} + \frac{1}{27} + \frac{1}{81} + \cdots + \frac{1}{3^k} + \cdots = \frac{1/3}{1 - 1/3} = \frac{1/3}{2/3} = \frac{1}{2}$$

To the mathematical sophisticate, this convergence argument presents a naive treatment of infinite series, for a modern development of this topic is considerably more subtle. This proof also obscures the reason why we initially assumed that $-1 < \alpha < 1$, although the need for this assumption is *suggested* by the geometric series with $\alpha = 2$. In this case, a direct application of our formula yields:

$$2 + 2^2 + 2^3 + 2^4 + \cdots = \frac{2}{1-2}$$

In other words, $2 + 4 + 8 + 16 + \cdots = -2$, a "double absurdity," both because the series clearly diverges to infinity and because, under no stretch of the imagination, could it add up to a *negative* number. The summation formula for geometric series, then, requires that α fall between -1 and 1. (A more detailed analysis of this matter is usually taken up in a course on calculus.)

The two previous infinite series illustrate an important condition on the general question of convergence. For the first geometric series with $\alpha = 1/3$, the succeeding terms we added on— $1/9$, $1/27$, $1/81$, etc.—were getting closer and closer to zero; hence the amount by which the sums grew for each additional term was getting ever more negligible. On the other hand, for the geometric series with $\alpha = 2$, we added terms that were moving away from zero—4, 8, 16, and so forth—and whose increasing magnitudes prevented the sum from homing in on a single number.

With these examples in mind, it would be quite reasonable to pose the following conjecture: an infinite series $x_1 + x_2 + x_3 + x_4 + \cdots + x_k + \cdots$ converges to a finite quantity *if and only if* the individual terms x_k themselves converge toward zero. As it turns out, half of this conjecture is true. That is, if the series converges to a finite sum, then the individual terms surely must get ever closer to zero. Put another way, we cannot hope to get an infinite number of terms to add to a finite sum unless the terms themselves become ever more negligible.

Unfortunately, the converse fails. That is, there are infinite series which, even though their individual terms melt away toward zero, have sums that grow toward infinity. This fact is by no means obvious, and it is precisely the content of the great theorem that follows. Upon examining the harmonic series $1 + 1/2 + 1/3 + 1/4 + 1/5 + \cdots + 1/k + \cdots$—that is, the sum of the reciprocals of the positive integers—Johann Bernoulli found that, whereas the individual terms clearly get closer and closer to zero, their sum nevertheless becomes infinite.

Bernoulli had discovered what mathematicians now call a "pathological counterexample"—that is, a specific example that seems so counterintuitive and bizarre as to warrant the label "pathological." Here is what is so unsettling about the harmonic series: one must add the first 83 terms of the series before its sum exceeds 5, since

$$1 + \frac{1}{2} + \frac{1}{3} + \frac{1}{4} + \frac{1}{5} + \cdots + \frac{1}{82} = 4.990020\ldots < 5.00 \quad \text{while}$$

$$1 + \frac{1}{2} + \frac{1}{3} + \frac{1}{4} + \frac{1}{5} + \cdots + \frac{1}{83} = 5.002068\ldots > 5.00$$

Note the remarkable fact that each and every term of the harmonic series *beyond* this one is itself less than $\frac{1}{83}$, and thus contributes very little to the overall sum. Consequently, we must add up another 144 terms just to get the sum to edge above 6. For this very slowly growing series to total 10, we have to sum the first 12,367 terms, and for it to climb to a sum of 20 takes somewhere around a quarter of a billion terms! To think that the harmonic series could eventually surpass one hundred, or one thousand, or one trillion seems completely out of the question.

But it does! This is what makes it pathological and what makes Bernoulli's theorem worthy of our attention.

Great Theorem: The Divergence of the Harmonic Series

Although this proof was concocted by Johann Bernoulli, it appeared in brother Jakob's 1689 *Tractatus de seriebus infinitis* (*Treatise on Infinite Series)*. With uncharacteristic fraternal affection, Jakob even prefaced the argument with an acknowledgment of his brother's priority.

Johann had to show that the harmonic series diverged to infinity. The proof rested on Leibniz's summation of the *convergent* series $\frac{1}{2} + \frac{1}{6} + \frac{1}{12} + \frac{1}{20} + \frac{1}{30} + \cdots = 1$, which we examined earlier in this chapter. This in itself is something of a surprise, for it is by no means clear how this well-behaved convergent series can shed light on the bizarre behavior of the harmonic series. Nonetheless, Johann Bernoulli reasoned as follows.

THEOREM The harmonic series $1 + \frac{1}{2} + \frac{1}{3} + \cdots + 1/k + \cdots$ is infinite.

PROOF Introduce $A = \frac{1}{2} + \frac{1}{3} + \frac{1}{4} + \frac{1}{5} + \cdots + 1/k + \cdots$, which is just the harmonic series lacking its first term. This series, ". . . transformed into fractions whose numerators are 1, 2, 3, 4, etc.," becomes

$$A = \frac{1}{2} + \frac{2}{6} + \frac{3}{12} + \frac{4}{20} + \frac{5}{30} + \cdots$$

which Johann noted for future reference.

Next, he designated by C the series of Leibniz already cited, and from this constructed a string of related series by subtracting in succession $\frac{1}{2}$, $\frac{1}{6}$, $\frac{1}{12}$, $\frac{1}{20}$, etc. This yielded

XVI. *Summa ſeriei infinitæ harmonicè progreſſionalium*, $\frac{1}{1}+\frac{1}{2}+\frac{1}{3}+\frac{1}{4}+\frac{1}{5}$ *&c. eſt infinita.*

Id primus deprehendit Frater: inventa namque per præced. ſumma ſeriei $\frac{1}{2}+\frac{1}{6}+\frac{1}{12}+\frac{1}{20}+\frac{1}{30}$, &c. viſurus porrò, quid emergeret ex iſta ſerie, $\frac{1}{2}+\frac{2}{6}+\frac{3}{12}+\frac{4}{20}+\frac{5}{30}$, &c. ſi reſolveretur methodo Prop. XIV. collegit propoſitionis veritatem ex abſurditate manifeſta, quæ ſequeretur, ſi ſumma ſeriei harmonicæ finita ſtatueretur. Animadvertit enim,

Seriem A, $\frac{1}{2}+\frac{1}{3}+\frac{1}{4}+\frac{1}{5}+\frac{1}{6}+\frac{1}{7}$, &c. ∞ (fractionibus ſingulis in alias, quarum numeratores ſunt 1, 2, 3, 4, &c. transmutatis) ſeriei B, $\frac{1}{2}+\frac{2}{6}+\frac{3}{12}+\frac{4}{20}+\frac{5}{30}+\frac{6}{42}$, &c. ∞ C+D+E+F, &c.

$$\left.\begin{array}{ll}
C.\ \frac{1}{2}+\frac{1}{6}+\frac{1}{12}+\frac{1}{20}+\frac{1}{30}+\frac{1}{42}, \&c. \infty \text{ per præc. } \frac{1}{1} & \\
D.\ .\ +\frac{1}{6}+\frac{1}{12}+\frac{1}{20}+\frac{1}{30}+\frac{1}{42}, \&c. \infty\ C-\frac{1}{2} \infty \frac{1}{2} & \\
E.\ .\ .\ +\frac{1}{12}+\frac{1}{20}+\frac{1}{30}+\frac{1}{42}, \&c. \infty\ D-\frac{1}{6} \infty \frac{1}{3} & \\
F.\ .\ .\ .\ .\ .\ +\frac{1}{20}+\frac{1}{30}+\frac{1}{42}, \&c. \infty\ E-\frac{1}{12} \infty \frac{1}{4} & \\
\qquad \&c. \infty \qquad \&c. &
\end{array}\right\} \infty\ G; \text{ unde sequitur,}$$

ſeriem G ∞ A, totum parti, ſi ſumma finita eſſet.

Ego

Johann's divergence proof, from Jakob's *Tractatus de seriebus infinitis*, republished in 1713 (photograph courtesy of The Ohio State University Libraries)

$$\begin{array}{llll}
C = \frac{1}{2}+\frac{1}{6}+\frac{1}{12}+\frac{1}{20}+\frac{1}{30}+\cdots & & & = 1 \\
D = \frac{1}{6}+\frac{1}{12}+\frac{1}{20}+\frac{1}{30}+\cdots & = C-\frac{1}{2} & = 1-\frac{1}{2} & = \frac{1}{2} \\
E = \frac{1}{12}+\frac{1}{20}+\frac{1}{30}+\cdots & = D-\frac{1}{6} & = \frac{1}{2}-\frac{1}{6} & = \frac{1}{3} \\
F = \frac{1}{20}+\frac{1}{30}+\cdots & = E-\frac{1}{12} & = \frac{1}{3}-\frac{1}{12} & = \frac{1}{4} \\
G = \frac{1}{30}+\cdots & = F-\frac{1}{20} & = \frac{1}{4}-\frac{1}{20} & = \frac{1}{5} \\
\vdots & \vdots & \vdots & \vdots
\end{array}$$

Johann now added down the two leftmost columns in this array of equations to get

$$\begin{aligned}
&C+D+E+F+\cdots \\
&\quad = \tfrac{1}{2}+\left(\tfrac{1}{6}+\tfrac{1}{6}\right)+\left(\tfrac{1}{12}+\tfrac{1}{12}+\tfrac{1}{12}\right)+\left(\tfrac{1}{20}+\tfrac{1}{20}+\tfrac{1}{20}+\tfrac{1}{20}\right)+\cdots \\
&\quad = \tfrac{1}{2}+\tfrac{2}{6}+\tfrac{3}{12}+\tfrac{4}{20}+\cdots = A \qquad \text{from above}
\end{aligned}$$

On the other hand, when summing down the leftmost and rightmost columns of the previous array, he found

$$C + D + E + F + G + \cdots = 1 + 1/2 + 1/3 + 1/4 + 1/5 + \cdots = 1 + A$$

Thus, since $C + D + E + F + G + \cdots$ equals both A and $1 + A$, Johann could only conclude that $1 + A = A$. As he put it, "the whole equals the part." But obviously no finite quantity is equal to one more than itself. To Johann Bernoulli, this could mean only one thing: that $1 + A$ is an infinite quantity. And, since $1 + A$ was the sum of the harmonic series, his argument was complete.

Q.E.D.

There are points about this proof that today's mathematicians can justly criticize. Bernoulli treated infinite series "holistically" as individual entities to be manipulated at will. We now know that a great deal more care is required when attacking these mathematical objects. Further, his technique of proving divergence contrasts dramatically with the modern approach. Today's mathematician would proceed as follows: fix a whole number N (no matter how large) and show that the series must exceed N; then, since it surpasses any whole number, the series must diverge to infinity. But Johann did no such thing. Instead, he proved divergence simply by showing that $A = 1 + A$, to modern tastes a most peculiar way of establishing that a positive quantity is infinite.

Before we become overly critical, we must acknowledge that Bernoulli was writing a century and a half before a truly rigorous theory of series was developed. Moreover, in spite of all objections, one cannot deny the sheer cleverness of Johann's argument. It was a mathematical gem.

At this point in the *Tractatus*, Jakob stressed the critical, and non-intuitive, consequence of his brother's proof by observing: "The sum of an infinite series whose final term vanishes perhaps is finite, perhaps infinite." Modern mathematicians cringe at his reference to the "final term" of an infinite series, for surely the nature of such series rules out any concluding term; however, his meaning is clear enough. He was highlighting the fact that, even if the individual terms of an infinite series approach zero, the sum of the series may yet be infinite. The harmonic series is the primary example of just this phenomenon, as Johann had shown.

Perhaps it was the unexpectedness of this result that moved Jakob to pen this bit of mathematical verse:

> As the finite encloses an infinite series
> And in the unlimited limits appear,

So the soul of immensity dwells in minutia
And in narrowest limits no limits inhere.
What joy to discern the minute in infinity!
The vast to perceive in the small, what divinity!

The Challenge of the Brachistochrone

With this and so many other contributions, the Bernoulli brothers left a significant mark upon the mathematics of their day. But one additional tale must be told of these cantankerous, competitive, and contentious brothers, a story that is surely one of the most fascinating from the entire history of mathematics.

It began in June of 1696 when Johann Bernoulli published a challenge problem in Leibniz's journal *Acta Eruditorum.* Obviously, a legacy of public challenges remained from the days of Fior and Tartaglia. Although the contests were now conducted in the sedate pages of scholarly journals, they retained their power to make or break reputations, as Johann himself observed:

> . . . it is known with certainty that there is scarcely anything which more greatly excites noble and ingenious spirits to labors which lead to the increase of knowledge than to propose difficult and at the same time useful problems through the solution of which, as by no other means, they may attain to fame and build for themselves eternal monuments among posterity.

Johann's particular challenge was a good one. He imagined points *A* and *B* at different heights above the ground and not lying one directly above the other. There is certainly an infinitude of different curves connecting these two points, from a straight line, to an arc of a circle, to any number of other wavy, undulating paths. Now imagine a ball rolling from *A* down to *B* along such a curve. The time it takes to complete the trip depends, of course, on the curve's shape. Bernoulli challenged the mathematical world to find that one particular curve *AMB* along which the ball will roll in the *shortest* time (see Figure 8.3). He called this curve the "brachistochrone" from the Greek words for "shortest" and "time."

An obvious first guess is to take *AMB* as the straight line joining *A* and *B*. But Johann cautioned against this simplistic approach:

> . . . to forestall hasty judgment, although the straight line *AB* is indeed the shortest between the points *A* and *B*, it nevertheless is not the path traversed in the shortest time. However the curve *AMB*, whose name I shall give if no

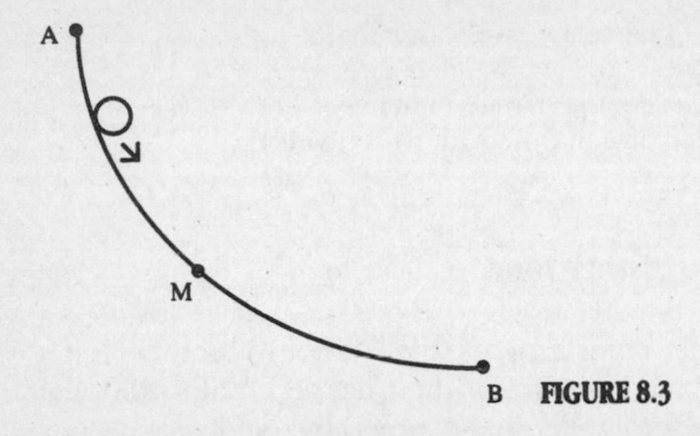

FIGURE 8.3

> one else has discovered it before the end of this year, is one well-known to geometers.

Johann gave the mathematical world until January 1, 1697, to come up with a solution. However, when his deadline arrived, he had received but one solution, from the "celebrated Leibniz," who

> has courteously asked me to extend the time limit to next Easter in order that in the interim the problem might be made public . . . that no one might have cause to complain of the shortness of time allotted. I have not only agreed to this commendable request but I have decided to announce myself the prolongation and shall now see who attacks this excellent and difficult question and after so long a time finally masters it.

Then, to be certain that no one misunderstood the problem, Johann repeated it:

> Among the infinitely many curves which join the two given points . . . choose the one such that, if the curve is replaced by a thin tube or groove, and a small sphere placed in it and released, then this will pass from one point to the other in the shortest time.

At this point, Johann waxed enthusiastic about the rewards of solving his brachistochrone problem. Recalling that he himself knew the solution, one finds his remarks about the glories of mathematics a bit self-serving:

> Let who can seize quickly the prize which we have promised to the solver. Admittedly this prize is neither of gold nor silver, for these appeal only to base and venal souls. . . . Rather, since virtue itself is its own most desirable reward and fame is a powerful incentive, we offer the prize, fitting for the man of noble blood, compounded of honor, praise, and approbation. . . .

In this passage, it sounds as though Johann was setting himself up for another triumph over poor Jakob. But he had a different target more squarely in his sights. Wrote Johann:

> . . . so few have appeared to solve our extraordinary problem, even among those who boast that through special methods . . . they have not only penetrated the deepest secrets of geometry but also extended its boundaries in marvellous fashion; although their golden theorems, which they imagine known to no one, have been published by others long before.

Could anyone doubt that the "golden theorems" he referred to were the techniques of fluxions, or that the object of his scorn was none other than Isaac Newton himself, a man who claimed to have known about calculus long before Leibniz had published it in 1684? Then, to leave no doubt about the explicit nature of his challenge, Johann put a copy in an envelope and mailed it off to England.

Of course, in 1697, Newton was deeply involved with matters of the Mint, and, as he himself admitted, he no longer felt the agility of mind that characterized his mathematical heyday. Newton was then living in London with his niece, Catherine Conduitt, and she picks up the story:

> When the problem in 1697 was sent by Bernoulli—Sir I. N. was in the midst of the hurry of the great recoinage and did not come home till four from the Tower very much tired, but did not sleep till he had solved it, which was by four in the morning.

Even late in life and tired from a hectic day's work, Isaac Newton triumphed where most of Europe had failed! It was a remarkable display of the powers of the great British genius. He had clearly felt his reputation and honor were on the line; after all, both Bernoulli and Leibniz were waiting in the wings to publish their own solutions. So Newton rose to the occasion and solved the problem in a matter of hours. Somewhat exasperated, he is reported at one point to have said, "I do not love . . . to be . . . teezed by forreigners about Mathematical things."

Back in Europe, as Easter neared, a few solutions came into the hands of Johann Bernoulli. The curve that everyone was seeking—one that "is well-known to geometers"—was none other than an upside-down cycloid. As we have noted, this important curve was studied by Pascal and Huygens, but neither of these mathematicians had realized that it would also serve as the curve of quickest descent. Johann wrote with characteristic hyperbole, ". . . you will be petrified with astonishment when I say that precisely this cycloid . . . of Huygens is our required brachistochrone."

On Easter, the challenge period had expired. All together, Johann had received five solutions. There was his own and the one from Leibniz. His brother Jakob came through (perhaps to Johann's dismay) with a third, and the Marquis de l'Hospital added a fourth. Finally, there was a submission bearing an English postmark. Opening it, Johann found the solution correct, although anonymous. He clearly had met his match in the person of Isaac Newton. Although unsigned, the solution bore the unmistakable signs of supreme genius.

There is a legend—probably of dubious authenticity but nonetheless of great charm—that Johann, partially chastened, partially in awe, put down the unsigned document and knowingly remarked, "I recognize the lion by his paw."

Epilogue

In describing Johann's proof of the divergence of the harmonic series, Jakob had begun, "*Id primus deprehendit frater*"—that is, "My brother discovered it first." If Jakob thought that Johann was the first to grasp the strange behavior of this series, he was quite wrong, for at least two earlier mathematicians had proved its divergence. Their proofs differed from each other's and from Johann's argument above, yet each exhibited its own special cleverness.

The earliest proof that the harmonic series diverges appeared in the work of Nicole Oresme (ca. 1323–1382), a French scholar of the fourteenth century. Oresme wrote a remarkable book with the title *Quaestiones super Geometriam Euclidis* sometime around 1350. This is, of course, a very old document, predating Cardano's *Ars Magna* by two full centuries. Yet, in spite of its origins in what we might call the "Stone Age" of European mathematics, Oresme's work contained some very nice results.

In particular, he addressed the nature of the harmonic series. Virtually his *entire* argument follows:

> . . . add to a magnitude of 1 foot: $\frac{1}{2}$, $\frac{1}{3}$, $\frac{1}{4}$ foot, etc.; the sum of which is infinite. In fact, it is possible to form an infinite number of groups of terms with a sum greater than $\frac{1}{2}$. Thus: $\frac{1}{3} + \frac{1}{4}$ is greater than $\frac{1}{2}$; $\frac{1}{5} + \frac{1}{6} + \frac{1}{7} + \frac{1}{8}$ is greater than $\frac{1}{2}$; $\frac{1}{9} + \frac{1}{10} + \cdots + \frac{1}{16}$ is greater than $\frac{1}{2}$, etc.

The reader may be forgiven for finding this a bit confusing. After all, the proof is entirely verbal, as one would expect for an argument written centuries before the appearance of symbolic algebra. Yet, with a bit of "cleaning up," it becomes a remarkably simple and clever divergence proof. His insight was to replace groups of fractions in the harmonic series by smaller fractions that sum to one-half. That is, he said:

$$1 + \tfrac{1}{2} > \tfrac{1}{2} + \tfrac{1}{2} = 1$$

$$1 + \tfrac{1}{2} + \left(\tfrac{1}{3} + \tfrac{1}{4}\right) > 1 + \left(\tfrac{1}{4} + \tfrac{1}{4}\right) = \tfrac{3}{2}$$

$$1 + \tfrac{1}{2} + \tfrac{1}{3} + \tfrac{1}{4} + \left(\tfrac{1}{5} + \tfrac{1}{6} + \tfrac{1}{7} + \tfrac{1}{8}\right)$$
$$> \tfrac{3}{2} + \left(\tfrac{1}{8} + \tfrac{1}{8} + \tfrac{1}{8} + \tfrac{1}{8}\right) = \tfrac{4}{2}$$

$$1 + \tfrac{1}{2} + \cdots + \tfrac{1}{8} + \left(\tfrac{1}{9} + \tfrac{1}{10} + \cdots + \tfrac{1}{16}\right)$$
$$> \tfrac{4}{2} + \left(\tfrac{1}{16} + \cdots + \tfrac{1}{16}\right) = \tfrac{5}{2}$$

This process can be extended, so that, in general, for any whole number k,

$$1 + \frac{1}{2} + \frac{1}{3} + \cdots + \frac{1}{2^k} > \frac{k+1}{2}$$

For instance, if $k = 9$, we see that

$$1 + \frac{1}{2} + \frac{1}{3} + \cdots + \frac{1}{512} = 1 + \frac{1}{2} + \frac{1}{3} + \cdots + \frac{1}{2^9} > \frac{9+1}{2} = 5$$

For $k = 99$, we get

$$1 + \frac{1}{2} + \frac{1}{3} + \cdots + \frac{1}{2^{99}} > \frac{100}{2} = 50$$

and for $k = 9999$ we have

$$1 + \frac{1}{2} + \frac{1}{3} + \cdots + \frac{1}{2^{9999}} > \frac{9999+1}{2} = 5000$$

So, by taking enough terms of the harmonic series, we can ensure that its sum exceeds 5, or 50, or 5000, or in general any finite quantity. This guarantees that the entire harmonic series, being greater than any finite quantity, will diverge to infinity. Oresme's proof, clever, concise and easily remembered, is the one found in most modern mathematics texts. The Bernoullis, however, seem not to have been aware of its existence.

Johann Bernoulli had been anticipated by yet another mathematician, the Italian Pietro Mengoli (1625–1686). Mengoli's argument dates from 1647, and thus predated the Bernoullian proof by four decades. It was quite a simple argument, provided one first established a preliminary result.

THEOREM If $a > 1$, then $\dfrac{1}{a-1} + \dfrac{1}{a} + \dfrac{1}{a+1} > \dfrac{3}{a}$.

PROOF Begin with the obvious fact that $2a^3 > 2a^3 - 2a = 2a(a^2 - 1)$ and divide both sides of this inequality by $a^2(a^2 - 1)$ to get

$$\frac{2a^3}{a^2(a^2-1)} > \frac{2a(a^2-1)}{a^2(a^2-1)} \quad \text{or simply} \quad \frac{2a}{a^2-1} > \frac{2}{a}$$

Thus

$$\begin{aligned}
\frac{1}{a-1} + \frac{1}{a} + \frac{1}{a+1} &= \frac{1}{a} + \left(\frac{1}{a-1} + \frac{1}{a+1}\right) \\
&= \frac{1}{a} + \frac{2a}{a^2-1} \qquad \text{by a bit of algebra} \\
&> \frac{1}{a} + \frac{2}{a} \qquad \text{by the inequality above} \\
&= \frac{3}{a}
\end{aligned}$$

Q.E.D.

This proposition guarantees that, when adding the reciprocals of three consecutive whole numbers, the sum must exceed three times the reciprocal of the middle number. For instance, as one can check numerically,

$$\tfrac{1}{8} + \tfrac{1}{9} + \tfrac{1}{10} > \tfrac{3}{9} = \tfrac{1}{3} \quad \text{or} \quad \tfrac{1}{32} + \tfrac{1}{33} + \tfrac{1}{34} > \tfrac{3}{33} = \tfrac{1}{11}$$

This was precisely the result that Mengoli needed to attack the harmonic series in his little proof from 1647.

THEOREM The harmonic series diverges to infinity.

PROOF Let H be the sum of the harmonic series. By grouping its terms and repeatedly applying the previous inequality, we find:

$$\begin{aligned}
H &= 1 + (1/2 + 1/3 + 1/4) + (1/5 + 1/6 + 1/7) + (1/8 + 1/9 + 1/10) \\
&\qquad + (1/11 + 1/12 + 1/13) + \cdots \\
&> 1 + (3/3) + (3/6) + (3/9) + (3/12) + (3/15) + \cdots \\
&= 1 + 1 + 1/2 + 1/3 + 1/4 + 1/5 + 1/6 + 1/7 + 1/8 + 1/9 + \cdots \\
&= 2 + (1/2 + 1/3 + 1/4) + (1/5 + 1/6 + 1/7) + (1/8 + 1/9 + 1/10) \\
&\qquad + (1/11 + 1/12 + 1/13) + \cdots \\
&> 2 + (3/3) + (3/6) + (3/9) + (3/12) + (3/15) + \cdots \\
&= 2 + 1 + 1/2 + 1/3 + 1/4 + 1/5 + 1/6 + 1/7 + 1/8 + 1/9 + \cdots \\
&= 3 + (1/2 + 1/3 + 1/4) + (1/5 + 1/6 + 1/7) + (1/8 + 1/9 + 1/10) \\
&\qquad + (1/11 + 1/12 + 1/13) + \cdots
\end{aligned}$$

and so on. The beauty of Mengoli's argument is its self-replicating nature. Each time he applied his preliminary theorem to the harmonic series, he encountered the harmonic series again, but this time augmented by one unit. As we look at the inequalities generated above, we see that H exceeds 1, exceeds 2, exceeds 3, and indeed, by repeating this process, exceeds any finite quantity at all. With Mengoli, we are left to conclude that the sum of the harmonic series must be infinite.

Q.E.D.

So, Johann's great theorem, although proved differently, had been discovered before him by both Oresme and Mengoli. Moreover, immediately after he gave Johann's proof in *Tractatus*, Jakob presented his *own* proof of the divergence of the harmonic series. It too was a very elegant argument—somewhat advanced to be presented here—although its presence smacked of the mathematical combat that raged between these sibling rivals.

In *Tractatus*, with the harmonic series behind him, Jakob next addressed the sum of the reciprocals of the *squares* of the whole numbers. That is, he investigated

$$1 + \frac{1}{2^2} + \frac{1}{3^2} + \cdots + \frac{1}{k^2} + \cdots = 1 + 1/4 + 1/9 + 1/16 + \cdots$$

He noticed that $1/4 < 1/3$, that $1/9 < 1/6$, that $1/16 < 1/10$, and in general

$$\frac{1}{k^2} < \frac{1}{k(k+1)/2}$$

Therefore, he reasoned that

$$1 + \frac{1}{4} + \frac{1}{9} + \frac{1}{16} + \cdots + \frac{1}{k^2} + \cdots < 1 + \frac{1}{3} + \frac{1}{6} + \frac{1}{10} + \frac{1}{15} + \cdots$$

$$+ \frac{2}{k(k+1)} + \cdots = 2(\tfrac{1}{2} + \tfrac{1}{6} + \tfrac{1}{12} + \tfrac{1}{20} + \tfrac{1}{30} + \cdots) = 2(1) = 2$$

where we have once again applied Leibniz's ubiquitous sum from the beginning of this chapter. In this fashion, Jakob had shown that the series in question converged to *some* finite quantity less than 2. This technique of establishing convergence is now called the "comparison test," for obvious reasons. Jakob's argument provided an early instance of the comparison test in action.

Although they knew that this infinite series converged, the Bernoullis were unsuccessful in determining the exact value of its sum. Jakob reported his failure with the somewhat desperate plea: "If anyone finds and communicates to us that which up to now has eluded our efforts, great will be our gratitude."

Evaluating the series $1 + \frac{1}{4} + \frac{1}{9} + \frac{1}{16} + \cdots$ proved to be a very difficult matter, and it would take a genius beyond that of the Bernoullis to determine its elusive sum.

Interestingly, the problem was finally solved in 1734 by a young man who had studied mathematics under Johann Bernoulli himself. In summing this series, as in so many other areas of mathematics, this young man would turn out to surpass his teacher. In fact, he would surpass virtually everyone who ever put pencil to paper in a mathematical quest. This student was Leonhard Euler, author of our next great theorem.

Chapter 9

The Extraordinary Sums of Leonhard Euler

(1734)

The Master of All Mathematical Trades

The legacy of Leonhard Euler (pronounced "oiler") is unsurpassed in the long history of mathematics. In both quantity and quality, his achievements are overwhelming. Euler's collected works fill over 70 large volumes, a testament to the genius of this unassuming Swiss citizen who changed the face of mathematics so profoundly. Indeed, one's first inclination, upon encountering the volume and quality of his work, is to regard his story as an exaggerated piece of fiction rather than hard historical fact.

This remarkable individual was born in Basel, Switzerland, in 1707. Not surprisingly, he showed signs of genius as a youth. Euler's father, a Calvinist preacher, managed to work out an arrangement whereby young Leonhard would study with the renowned Johann Bernoulli. Euler later recalled these sessions with the master. The boy would work throughout the week and then, during an appointed hour on Saturday afternoons, would ask Bernoulli for help on the mathematical topics that had eluded him. Bernoulli, not always the most kind-hearted of men, may have initially shown some irritation at the shortcomings of his pupil; Euler, for his part, resolved to work as diligently as possible so as not to bother his mentor with unnecessary trifles.

Grumpy or not, Johann Bernoulli soon recognized the talent he was nurturing. Soon Euler was publishing mathematical papers of high quality, and at age 19 he won a prize from the French Academy for his brilliant analysis of the optimum placement of masts on a ship. (It should be noted that, at this point in his life, Euler had never even *seen* an ocean-going vessel!)

In 1727, Euler was appointed to the St. Petersburg Academy in Russia. At that time, the Russian establishment was trying to implement Peter the Great's dream of building an institution to rival the great academies of Paris and Berlin. Among the scholars lured to Russia was Daniel Bernoulli, son of Johann, and it was through Daniel's influence that Euler obtained his employment. Oddly, with the positions in the natural sciences being filled, Euler's appointment was in the areas of medicine and physiology. But a position was a position, so Euler readily accepted. After a shaky start, including a peculiar stint as a medical officer in the Russian navy, Euler at last landed a mathematical chair in 1733 when Daniel Bernoulli, its previous occupant, vacated it to return to Switzerland.

By then, Euler had already displayed the boundless energy and enormous creativity that would characterize his mathematical life. Although Euler began losing sight in his right eye during the mid-1730s and soon was virtually sightless in that eye, this physical impairment had no impact whatever on his scientific work. He continued unimpeded, solving significant problems from such diverse mathematical arenas as geometry, number theory, and combinatorics, as well as applied areas such as mechanics, hydrodynamics, and optics. It is both poignant and somehow remarkably uplifting to imagine a man slipping into blindness yet explaining to the world the mysteries of optical light.

In 1741, Euler left St. Petersburg to take a position in the Berlin Academy under Frederick the Great. This move was based in part upon his distaste for the repressive nature of the Czarist system. Unfortunately, the Berlin situation also proved to be far from ideal. Frederick regarded Euler as too little the sophisticate, too much the quiet, unassuming scholar. The German king, in an insensitive reference to Euler's vision problems, called him a "mathematical cyclops." Such treatment, along with the petty controversies and political in-fighting of the Academy, led Euler back to St. Petersburg during the reign of Catherine the Great, and he remained there until his death 17 years later.

Euler was described by contemporaries as a kind and generous man, one who enjoyed the simple pleasures of growing vegetables and telling

Leonhard Euler (photograph courtesy of The Ohio State University Libraries)

stories to his brood of 13 children. In this regard, Euler presents a welcome contrast to the withdrawn, secretive Isaac Newton, one of his very few mathematical peers. It is comforting to know that genius of this order does not necessarily bring with it a neurotic personality. Euler even retained his good nature when, in 1771, he lost most of the vision in his normal eye. Almost totally blind and in some pain, Euler nonetheless continued his mathematical writings unabated, by dictating his wonderful equations and formulas to an associate. Just as deafness proved no obstacle to Ludwig von Beethoven a generation later, so blindness did not reduce the flow of mathematics from Leonhard Euler.

Throughout his career, Euler was blessed with a memory that can only be called phenomenal. His number-theoretic investigations were aided by the fact that he had memorized not only the first 100 prime numbers but also all of their squares, their cubes, and their fourth, fifth, and sixth powers. While others were digging through tables or pulling out pencil and paper, Euler could simply recite from memory such quantities as 241^4 or 337^6. But this was the least of his achievements. He was able to do difficult calculations mentally, some of these requiring him to retain in his head up to 50 places of accuracy! The Frenchman Francois Arago said that Euler calculated without apparent effort, "just as men breathe, as eagles sustain themselves in the air." Yet this extraordinary mind still had room for a vast collection of memorized facts, orations, and poems, including the entire text of Virgil's *Aeneid*, which Euler had committed to memory as a boy and still could recite flawlessly half a century later. No writer of fiction would dare to provide a character with a memory of this caliber.

Part of Euler's well-deserved reputation rests upon the textbooks he authored. While some of these were written at the highest level of mathematical sophistication, he did not find it demeaning to write more elementary books as well. Perhaps his best-known text was the *Introductio in Analysin Infinitorum* of 1748. This classic mathematical exposition has been compared to Euclid's *Elements* in that it surveyed the discoveries of earlier mathematicians, organized and cleaned up the proofs, and did the job so well as to render most previous writings obsolete. To the *Introductio* he added a volume on differential calculus in 1755 and three volumes on integral calculus in 1768–74, thereby charting the general direction for mathematical analysis down to the present day.

In all of his texts, Euler's exposition was quite lucid, and his mathematical notation was chosen so as to clarify, not obscure, the underlying ideas. Indeed, Euler's mathematical writings are the first that look truly modern to today's reader; this, of course, is not because he chose a modern notation but because his influence was so pervasive that all subsequent mathematicians adopted his style, notation, and format. Moreover, he wrote with an understanding that not all his readers had his awesome ability for learning mathematics. Euler was not the stereotypical mathematician who sees deeply into the nature of his subject but finds it impossible to convey his ideas to others. On the contrary, he cared deeply about teaching. Condorcet, in a wonderful little phrase, said of Euler: "He preferred instructing his pupils to the little satisfaction of amazing them." This is quite a compliment to a person who, if he had so chosen, could surely have amazed anyone with his mathematical prowess.

Any discussion of Euler's mathematics somehow returns to his *Opera Omnia*, those 73 volumes of collected papers. These contain the 886 books and articles—written variously in Latin, French, and German—that he produced during his career. His output was so huge and the pace of its production so rapid—even in the darkness of his later life—that a publication backlog is reported to have lasted 47 years after his death.

As noted, Euler did not confine his work to pure mathematics. Rather, his opus contains papers on acoustics, engineering, mechanics, astronomy, and even a three-volume treatise on optical devices such as telescopes and microscopes. Incredible as it sounds, it has been estimated that, if one were to collect *all* publications in the mathematical sciences produced over the last three-quarters of the eighteenth century, roughly one-third of these were from the pen of Leonhard Euler!

Standing in a library before his collected works, one surveys shelf upon shelf of large volumes with a sense of disbelief. Contained in those thousands of pages are seminal papers that charted new directions for whole areas of mathematics, from the calculus of variations, to graph theory, to complex analysis, to differential equations. Virtually every branch of mathematics has theorems of major significance that are attributed to Euler. Thus, we find the Euler triangle in geometry, the Euler characteristic in topology, and the Euler circuit in graph theory, not to mention such entities as the Euler constant, the Euler polynomials, the Euler integrals, and so on. And even this is but half the story, for a large number of mathematical results traditionally attributed to others were in fact discovered by Euler and appear neatly tucked away amid the huge body of his work. One wag has noted, not entirely in jest, that

> . . . there is ample precedent for naming laws and theorems for persons other than their discoverers, else half of analysis would be named for Euler.

Leonhard Euler died suddenly on September 7, 1783. Up until the end, he had been mathematically active, in spite of his blindness. Reportedly, he spent his last day playing with his grandchildren and discussing the latest theories about the planet Uranus. For Euler, the end came quickly when, in Condorcet's phrase, "He ceased to calculate and to live." He is buried in St. Petersburg (now called Leningrad), which had been his home, on and off, for so many happy years.

Great Theorem: Evaluating $1 + \frac{1}{4} + \frac{1}{9} + \frac{1}{16} + \frac{1}{25} + \cdots + \frac{1}{k^2} + \cdots$

From this enormous body of mathematics, it is difficult to select one or two representative theorems. That which follows was chosen for a number of reasons. First, its history renders it an important and provocative result. Second, it was one of Euler's early triumphs, announced in 1734 during his first years at St. Petersburg; by all accounts, it did much to solidify his reputation for mathematical genius. Finally, it provides an illustration not only of Euler's brilliance in solving a problem that had stumped his predecessors, but also of his ability to turn an individual solution into a string of equally impressive and utterly unexpected ones. While no single theorem can encompass the genius of Leonhard Euler, the one we shall now examine illustrates his powers clearly.

The problem is that with which we concluded Chapter 8. Recall that the Bernoullis, fresh from their triumph on the harmonic series, had examined the series $1 + 1/4 + 1/9 + 1/16 + 1/25 + \cdots$. While they knew it summed to a number less than 2, they had no idea what this sum was. Apparently, the evaluation of this series mocked not only Jakob and Johann Bernoulli, but even Leibniz himself, not to mention the rest of the world's mathematical community.

Euler would certainly have heard of this problem through his teacher Johann. He reported that his first attack on this series was simply to add up more and more of its terms in the hope of recognizing the sum. He carried this approximate sum to 20 places—no mean computational feat in the days before computers—and found that it was tending toward the number 1.6449. Unfortunately, this number did not look at all familiar. Not to be deterred, he kept up his attack on the problem until finally he discovered the key to unlock its mystery. With obvious enthusiasm he wrote, ". . . quite unexpectedly I have found an elegant formula . . . depending on π."

To derive this elegant formula, Euler needed two tools. One was the so-called "sine function" from elementary trigonometry. A full discussion of this important mathematical concept—usually written as

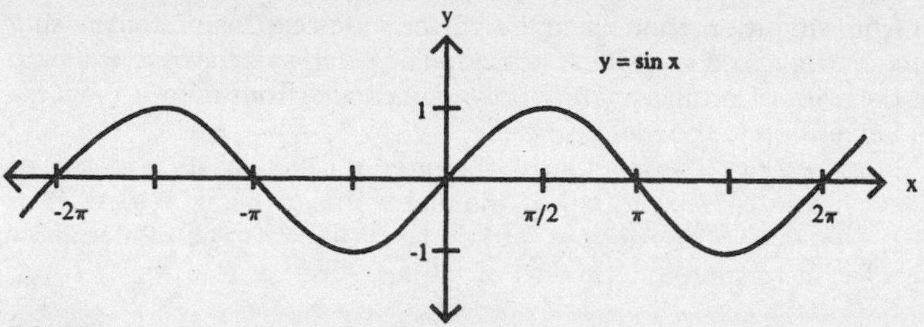

FIGURE 9.1

"sin x"—would carry us too far afield. However, anyone who has seen trigonometry or precalculus surely has met the well-known sine wave with its infinitely oscillating behavior. A graph of the function $f(x) = \sin x$ appears in Figure 9.1 and lay at the heart of Euler's insight.

Recalling that a function equals zero for precisely those values of x at which its graph cuts across the x-axis, we see that $\sin x = 0$ for $x = 0, \pm\pi, \pm 2\pi, \pm 3\pi, \pm 4\pi$, and so on. This infinite collection of x-values at which $\sin x$ equals zero reflects the repeating, periodic behavior of the sine function.

This much information about the sine can be garnered from a basic trigonometry course. But if we add to our arsenal the power of calculus, the following formula emerges:

$$\sin x = x - \frac{x^3}{3!} + \frac{x^5}{5!} - \frac{x^7}{7!} + \frac{x^9}{9!} - \cdots$$

Again, it is not necessary to go through the derivation of the formula, although those who have studied Taylor series expansions in calculus will recognize it at once. Its importance was in providing Euler with a representation of $\sin x$ as an "infinitely long polynomial."

This series for $\sin x$ needs a few words of explanation. First, the denominators involve the factorial notation common in certain branches of mathematics. By definition, $3!$ means $3 \times 2 \times 1 = 6$; $5! = 5 \times 4 \times 3 \times 2 \times 1 = 120$; and so on. Moreover, the expression for $\sin x$ will continue forever, with the powers on the x running through the sequence of odd integers, the denominators being the associated factorials, and the signs alternating between positive and negative. This is what we mean when we say we have written the $\sin x$ as an infinitely long polynomial. This was one of the clues Euler needed to solve the puzzle at hand.

The other fact came not from the domain of trigonometry or calculus, but from simple algebra. Since the Taylor series expansion for the sine function suggested an endless polynomial, Euler was drawn to examine the behavior of ordinary, *finite* polynomials and from there to make a bold extension to the infinite case.

Suppose $P(x)$ is a polynomial of degree n having as its n roots $x = a$, $x = b$, $x = c, \ldots,$ and $x = d$; in other words, $P(a) = P(b) = P(c) = \cdots = P(d) = 0$. Suppose further that $P(0) = 1$. Then Euler knew that $P(x)$ factors into the product of n linear terms as follows:

$$P(x) = \left(1 - \frac{x}{a}\right)\left(1 - \frac{x}{b}\right)\left((1 - \frac{x}{c}\right)\cdots\left(1 - \frac{x}{d}\right)$$

It may be wise to look at the reasonableness of this general formula. We can see by direct substitution that

$$P(a) = \left(1 - \frac{a}{a}\right)\left(1 - \frac{a}{b}\right)\left(1 - \frac{a}{c}\right)\cdots\left(1 - \frac{a}{d}\right) = 0$$

since the first factor is just $1 - 1 = 0$. Likewise,

$$P(b) = \left(1 - \frac{b}{a}\right)\left(1 - \frac{b}{b}\right)\left(1 - \frac{b}{c}\right)\cdots\left(1 - \frac{b}{d}\right) = 0$$

since now the second factor is $1 - 1 = 0$. Indeed, the expression for $P(x)$ shows quite clearly that $P(a) = P(b) = P(c) = \cdots = P(d) = 0$, as was desired.

But there was another condition upon $P(x)$: we demanded that $P(0) = 1$. Fortunately, our formula comes through here as well, for

$$P(0) = \left(1 - \frac{0}{a}\right)\left(1 - \frac{0}{b}\right)\left(1 - \frac{0}{c}\right)\cdots\left(1 - \frac{0}{d}\right)$$
$$= (1)(1)(1)\ldots(1) = 1$$

In short

$$P(x) = \left(1 - \frac{x}{a}\right)\left(1 - \frac{x}{b}\right)\left(1 - \frac{x}{c}\right)\cdots\left(1 - \frac{x}{d}\right)$$

has the properties we sought.

As an example, suppose that $P(x)$ is a cubic polynomial for which $P(2) = P(3) = P(6) = 0$ and $P(0) = 1$. Then we get the factorization

$$P(x) = \left(1 - \frac{x}{2}\right)\left(1 - \frac{x}{3}\right)\left(1 - \frac{x}{6}\right) = 1 - x + \frac{11}{36}x^2 - \frac{1}{36}x^3$$

and it can be easily checked that this cubic meets our conditions.

Euler, contemplating this equation, decided that a similar rule would surely hold for "infinite polynomials." Like Newton, he was a great believer in the persistence of patterns, and if the pattern was valid for the finite case, why not extend it to the infinite one? Modern mathematicians know that this can be a dangerous practice, and the extension of formulas from the finite to the infinite can lead to enormous difficulties. Such an extension certainly is more subtle, and demands more care, than Euler gave it. Perhaps he was just lucky; perhaps his mathematical intuition was particularly strong. In any case, his bold extension paid off.

These preliminary results may seem rather far removed from the initial problem of summing $1 + 1/4 + 1/9 + 1/16 + 1/25 + \cdots$. It took an insight of "Eulerian" proportions—which is to say an insight so profound as to be almost breathtaking—to tie all of the pieces together.

THEOREM $$1 + \frac{1}{4} + \frac{1}{9} + \frac{1}{16} + \frac{1}{25} + \cdots + \frac{1}{k^2} + \cdots = \frac{\pi^2}{6}$$

PROOF Euler began by introducing the function

$$f(x) = 1 - \frac{x^2}{3!} + \frac{x^4}{5!} - \frac{x^6}{7!} + \frac{x^8}{9!} - \cdots$$

To Euler, $f(x)$ was just an infinite polynomial with $f(0) = 1$ (as is immediately apparent). Thus, it can be factored, in the manner developed above, provided we determine the roots of the equation $f(x) = 0$. To this end, observe that, for $x \neq 0$

$$f(x) = x\left[\frac{1 - \frac{x^2}{3!} + \frac{x^4}{5!} - \frac{x^6}{7!} + \frac{x^8}{9!} - \cdots}{x}\right]$$

$$= \frac{x - \frac{x^3}{3!} + \frac{x^5}{5!} - \frac{x^7}{7!} + \frac{x^9}{9!} - \cdots}{x}$$

$$= \frac{\sin x}{x} \quad \text{by the Taylor Expansion of } \sin x$$

Therefore, so long as x is not 0, solving $f(x) = 0$ amounts to solving $\frac{\sin x}{x} = 0$, which (through a simple cross-multiplication) reduces to solving $\sin x = 0$. As we have seen, the sine function equals 0 precisely for $x = 0$, $x = \pm\pi$, $x = \pm 2\pi$, and so on. But we must, of course, eliminate $x = 0$ from contention as a solution to $f(x) = 0$, since we have already noted that $f(0) = 1$. In other words, the solutions of $f(x) = 0$ are just $x = \pm\pi, x = \pm 2\pi, x = \pm 3\pi, \ldots$

With these considerations behind him, Euler factored $f(x)$ as:

$$f(x) = \left(1 - \frac{x}{\pi}\right)\left(1 - \frac{x}{-\pi}\right)\left(1 - \frac{x}{2\pi}\right)\left(1 - \frac{x}{-2\pi}\right)\left(1 - \frac{x}{3\pi}\right)\left(1 - \frac{x}{-3\pi}\right)\cdots$$

$$= \left[\left(1 - \frac{x}{\pi}\right)\left(1 + \frac{x}{\pi}\right)\right]\left[\left(1 - \frac{x}{2\pi}\right)\left(1 + \frac{x}{2\pi}\right)\right]\left[\left(1 - \frac{x}{3\pi}\right)\left(1 + \frac{x}{3\pi}\right)\right]\cdots$$

which amounts to

$$1 - \frac{x^2}{3!} + \frac{x^4}{5!} - \frac{x^6}{7!} + \frac{x^8}{9!} - \cdots = \left[1 - \frac{x^2}{\pi^2}\right]\left[1 - \frac{x^2}{4\pi^2}\right]\left[1 - \frac{x^2}{9\pi^2}\right]\left[1 - \frac{x^2}{16\pi^2}\right]\cdots$$

We shall call this the key equation. It is a most extraordinary result, for it equates an infinite *sum* with an infinite *product*. That is, the infinite series by which $f(x)$ was originally defined has been equated to the infinite product on the right. To a mathematician of Euler's vision, this was highly suggestive. In fact, he was now on the brink of completing his proof, although many readers may still be completely in the dark as to where his argument is leading.

What Euler did was to imagine "multiplying out" the infinite product on the right side of the preceding equation and then collecting all terms having the same power of x. In so doing, the first term to appear would be the product of all of the 1s and this, of course, is 1. To end up with a term in x^2, we would have to multiply the 1s from all but one of the factors by an x^2 term from that remaining factor. Hence, Euler's "infinite multiplication" problem would result in the equation:

$$1 - \frac{x^2}{3!} + \frac{x^4}{5!} - \frac{x^6}{7!} + \frac{x^8}{9!} - \cdots$$
$$= \left[1 - \frac{x^2}{\pi^2}\right]\left[1 - \frac{x^2}{4\pi^2}\right]\left[1 - \frac{x^2}{9\pi^2}\right]\left[1 - \frac{x^2}{16\pi^2}\right]\cdots$$
$$= 1 - \left(\frac{1}{\pi^2} + \frac{1}{4\pi^2} + \frac{1}{9\pi^2} + \frac{1}{16\pi^2} + \cdots\right)x^2 + (\ldots)\,x^4 - \cdots$$

At last, the smoke begins to clear. That is, once Euler had multiplied out the infinite product to get *two* infinite sums equaling each other, nothing would be more natural than to equate the like powers of x. Note that both series begin with 1. Next comes the x^2 term in each series, and so their coefficients must be equal. That is,

$$-\frac{1}{3!} = -\left(\frac{1}{\pi^2} + \frac{1}{4\pi^2} + \frac{1}{9\pi^2} + \frac{1}{16\pi^2} + \cdots\right)$$

Then, multiplying both sides by -1, observing that $3! = 6$ on the left side, and factoring the common π^2 out of the right side, Euler arrived at

$$\frac{1}{6} = \frac{1}{\pi^2}\left(1 + \frac{1}{4} + \frac{1}{9} + \frac{1}{16} + \cdots\right)$$

and a final cross-multiplication yielded the astounding fact that

$$1 + \frac{1}{4} + \frac{1}{9} + \frac{1}{16} + \cdots = \frac{\pi^2}{6}$$

Q.E.D.

Here, then, Leonhard Euler found the answer that had escaped mathematicians for decades. Sure enough, the numerical value of $\pi^2/6$ is 1.6449 . . . , the approximate value Euler had originally determined. Note also that this sum is indeed less than 2, as Jakob Bernoulli had correctly deduced in 1689.

But no one before Euler had the slightest inkling that the series summed to exactly one-sixth of π^2. What a bizarre result this is. For reasons buried deeply within the mathematics itself, the sum of the series in question turns out to be, of all things, a formula involving π. Since π is naturally associated with circles, but since numbers like 1, 4, 9, 16 arise in conjunction with squares, an outcome linking the two could hardly have been anticipated. Even Euler was surprised by the answer. His formula was, and to this day remains, one of the most peculiar and surprising in all of mathematics. This unexpectedness, combined with the very clever method of deriving it, makes Euler's proof a great theorem of the first rank.

Epilogue

The result just presented helped establish the reputation of Leonhard Euler throughout the mathematical world. It was an undisputed triumph, and surely many lesser intellects would have been perfectly happy to sit back and rest on these very impressive laurels—but not Euler. Typical of his mathematics was the ability to mine any result for all it was worth. In the case of his wonderful sums, he had only scratched the surface.

Euler returned to the fact that the infinite product he had generated in the key equation was equal to $\sin x/x$, provided $x \neq 0$. Looking at the graph of the sine function in Figure 9.1, one sees that $\sin x$ reaches a peak of 1 when $x = \pi/2$. So, upon substituting $x = \pi/2$ into the infinite product, we find that:

$$\frac{\sin \pi/2}{\pi/2} = \left[1 - \frac{(\pi/2)^2}{\pi^2}\right]\left[1 - \frac{(\pi/2)^2}{4\pi^2}\right]\left[1 - \frac{(\pi/2)^2}{9\pi^2}\right]\left[1 - \frac{(\pi/2)^2}{16\pi^2}\right]\cdots$$

which became

$$\frac{1}{\pi/2} = \left(1 - \frac{1}{4}\right)\left(1 - \frac{1}{16}\right)\left(1 - \frac{1}{36}\right)\left(1 - \frac{1}{64}\right)\cdots$$

or simply

$$\frac{2}{\pi} = \left(\frac{3}{4}\right)\left(\frac{15}{16}\right)\left(\frac{35}{36}\right)\left(\frac{63}{64}\right)\cdots$$

Upon taking reciprocals and factoring the right side, Euler had stumbled upon the formula:

$$\frac{\pi}{2} = \frac{2 \times 2 \times 4 \times 4 \times 6 \times 6 \times 8 \times 8 \ldots}{1 \times 3 \times 3 \times 5 \times 5 \times 7 \times 7 \times 9 \ldots}$$

This expression, which gives $\pi/2$ as a huge quotient having the product of the even numbers in the numerator and the product of the odd ones in the denominator, had been known to the Englishman John Wallis (1616–1703), who derived it in a very different fashion as early as 1650. Thus, Euler had not so much discovered a new formula here as he had rediscovered it by his novel and obviously quite powerful use of infinite sums and infinite products.

But Euler had more tricks up his sleeve. He realized that the technique he developed to evaluate the series $1 + \frac{1}{4} + \frac{1}{9} + \frac{1}{16} + \cdots$ was the key to evaluating even "wilder" series. For instance, suppose we wanted to sum the reciprocals of just the *even* squares:

$$\frac{1}{4} + \frac{1}{16} + \frac{1}{36} + \frac{1}{64} + \frac{1}{100} + \cdots + \frac{1}{(2k)^2} + \cdots$$

Euler simply factored out $\frac{1}{4}$ and then referred to the "great theorem" to see that

$$\frac{1}{4} + \frac{1}{16} + \frac{1}{36} + \frac{1}{64} + \frac{1}{100} + \cdots$$
$$= \frac{1}{4}\left(1 + \frac{1}{4} + \frac{1}{9} + \frac{1}{16} + \frac{1}{25} + \cdots\right) = \frac{1}{4}\left(\frac{\pi^2}{6}\right) = \frac{\pi^2}{24}$$

Then, Euler had no trouble summing the reciprocals of the *odd* perfect squares as well, since

$$1 + \frac{1}{9} + \frac{1}{25} + \frac{1}{49} + \cdots = \left(1 + \frac{1}{4} + \frac{1}{9} + \frac{1}{16} + \frac{1}{25} + \cdots\right)$$
$$- \left(\frac{1}{4} + \frac{1}{16} + \frac{1}{36} + \frac{1}{64} + \frac{1}{100} + \cdots\right) = \frac{\pi^2}{6} - \frac{\pi^2}{24} = \frac{\pi^2}{8}$$

And still Euler pushed on, obviously exhilarated by his discoveries. He raised the question of summing the reciprocals of the *fourth powers* of the integers:

$$1 + \frac{1}{16} + \frac{1}{81} + \frac{1}{256} + \frac{1}{625} + \frac{1}{1296} + \cdots + \frac{1}{k^4} + \cdots$$

Could he determine this one?

Euler realized that he should return to the key equation and this time determine the coefficients of x^4 from each side of the equality. But how does one go about finding the x^4 term that emerges from the infinite multiplication on the right side of the key equation? This was no trivial question. In answering it, Euler was again aided by his keen sense of pattern recognition and his faith that whatever holds for finite products can be safely extended to infinite products.

In order to understand his reasoning, we shall first consider two simple but suggestive examples, keeping an eye on the resulting coefficients of x^4. First of all

$$\begin{aligned}(1 - ax^2)(1 - bx^2) &= 1 - (a + b)x^2 + abx^4 \\ &= 1 - (a + b)x^2 + \tfrac{1}{2}[(a + b)^2 - (a^2 + b^2)]x^4\end{aligned}$$

Secondly

$$\begin{aligned}&(1 - ax^2)(1 - bx^2)(1 - cx^2) \\ &\quad = 1 - (a + b + c)x^2 + (ab + ac + bc)x^4 - (abc)x^6 \\ &\quad = 1 - (a + b + c)x^2 + \tfrac{1}{2}[(a + b + c)^2 - (a^2 + b^2 + c^2)]x^4 \\ &\qquad - (abc)x^6\end{aligned}$$

These equations can be checked directly simply by multiplying out the terms in square brackets on the right-hand sides.

Note that a pattern has emerged—namely, when multiplying a series of factors like $(1 - ax^2)$, $(1 - bx^2)$, $(1 - cx^2)$, and so on, the coefficient of x^4 is half the difference between the square of the sum $(a + b + c + \cdots)$ and the sum of the squares $(a^2 + b^2 + c^2 + \cdots)$. If this pattern holds for the product of two or three such factors, why not extend it to four, or five, or to an infinite number? Returning to the key equation, Euler enthusiastically did just that:

$$\begin{aligned}&1 - \frac{x^2}{3!} + \frac{x^4}{5!} - \frac{x^6}{7!} + \frac{x^8}{9!} - \cdots \\ &\qquad = \left[1 - \frac{x^2}{\pi^2}\right]\left[1 - \frac{x^2}{4\pi^2}\right]\left[1 - \frac{x^2}{9\pi^2}\right]\left[1 - \frac{x^2}{16\pi^2}\right]\cdots\end{aligned}$$

Here we see that $1/\pi^2$ corresponds to a, $1/4\pi^2$ corresponds to b, $1/9\pi^2$ corresponds to c, and so on. Applying our observation about the coefficient of x^4 thus yields:

$$\begin{aligned}1 - \frac{x^2}{3!} + \frac{x^4}{5!} - \frac{x^6}{7!} + \frac{x^8}{9!} - \cdots &= 1 - \left(\frac{1}{\pi^2} + \frac{1}{4\pi^2} + \frac{1}{9\pi^2} + \cdots\right)x^2 \\ &\quad + \frac{1}{2}\left[\left(\frac{1}{\pi^2} + \frac{1}{4\pi^2} + \frac{1}{9\pi^2} + \frac{1}{16\pi^2} + \cdots\right)^2\right. \\ &\qquad \left. - \left(\frac{1}{\pi^4} + \frac{1}{16\pi^4} + \frac{1}{81\pi^4} + \frac{1}{256\pi^4} + \cdots\right)\right] x^4 - \cdots\end{aligned}$$

Euler now considered the coefficients of x^4 on both sides of this equation. The coefficient of x^4 on the left side is just $1/5! = 1/120$. The corresponding coefficient in the right-hand series is much more intricate but can be tidied up algebraically by factoring out common powers of π and using the conclusion of the great theorem above. That is, the coefficient of x^4 on the right is

$$\frac{1}{2}\left[\left(\frac{1}{\pi^2}+\frac{1}{4\pi^2}+\frac{1}{9\pi^2}+\frac{1}{16\pi^2}+\cdots\right)^2 - \left(\frac{1}{\pi^4}+\frac{1}{16\pi^4}+\frac{1}{81\pi^4}+\frac{1}{256\pi^4}+\cdots\right)\right]$$

$$=\frac{1}{2}\left[\frac{1}{\pi^4}\left(1+\frac{1}{4}+\frac{1}{9}+\frac{1}{16}+\cdots\right)^2 - \frac{1}{\pi^4}\left(1+\frac{1}{16}+\frac{1}{81}+\frac{1}{256}+\cdots\right)\right]$$

$$=\frac{1}{2\pi^4}\left[\left(\frac{\pi^2}{6}\right)^2-\left(1+\frac{1}{16}+\frac{1}{81}+\frac{1}{256}+\cdots\right)\right]$$

$$=\frac{1}{72}-\frac{1}{2\pi^4}\left(1+\frac{1}{16}+\frac{1}{81}+\frac{1}{256}+\cdots\right)$$

Now the moment of truth was at hand. Euler simply *equated* these coefficients of x^4 and solved the resulting equation as shown below:

$$\frac{1}{120}=\frac{1}{72}-\frac{1}{2\pi^4}\left(1+\frac{1}{16}+\frac{1}{81}+\frac{1}{256}+\cdots\right)$$

so

$$\frac{1}{2\pi^4}\left(1+\frac{1}{16}+\frac{1}{81}+\frac{1}{256}+\cdots\right)=\frac{1}{72}-\frac{1}{120}=\frac{1}{180}$$

Then a final cross multiplication yielded Euler's formula:

$$1+\frac{1}{16}+\frac{1}{81}+\frac{1}{256}+\cdots+\frac{1}{k^4}+\cdots=\frac{2\pi^4}{180}=\frac{\pi^4}{90}$$

Here Euler had discovered a genuinely odd result linking the reciprocals of the perfect fourth powers with the fourth power of π. Then, like a child with a newly found plaything, he exuberantly applied his remarkable techniques to sum even stranger series like:

$$1 + \frac{1}{64} + \frac{1}{729} + \frac{1}{4096} + \cdots + \frac{1}{k^6} + \cdots = \frac{\pi^6}{945}$$

and he kept at it for even exponents all the way to the phenomenal

$$1 + \frac{1}{2^{26}} + \frac{1}{3^{26}} + \cdots + \frac{1}{k^{26}} + \cdots = \frac{1315862}{11094481976030578125}\pi^{26}$$

when even *he* finally tired of this. Needless to say, no one in history had ever traveled these mathematical paths. No matter how frivolous they may appear from a practical standpoint, they were surely advances in human knowledge, discoveries of previously unsuspected relationships involving reciprocals of powers of whole numbers and that most important of constants, π.

A question immediately comes to mind: What is the sum of the reciprocals of the *odd* powers of the integers? For instance, can we evaluate the infinite series

$$1 + \frac{1}{2^3} + \frac{1}{3^3} + \frac{1}{4^3} + \cdots = 1 + \frac{1}{8} + \frac{1}{27} + \frac{1}{64} + \cdots$$

as well? Here even Euler was mute, and the past 200 years of mathematical research have advanced our knowledge of such odd powers very little. It is easy to conjecture that the sum in question is of the form $(p/q)\,\pi^3$ for some fraction p/q, but to this day no one knows if this conjecture is valid.

Today, we recognize that Euler was not so precise in his use of the infinite as he should have been. His belief that finitely generated patterns and formulas automatically extend to the infinite case was more a matter of faith than of science, and subsequent mathematicians would provide scores of examples showing the folly of such hasty generalizations. In short, Euler could be convicted of giving insufficient attention to the logical foundations of his arguments. Yet such criticisms barely tarnish his reputation. Even though his approach to infinite series was naive, all of these wonderful sums have been subsequently verified by today's higher standards of logical rigor.

These triumphs occupied just a few pages of the 70-odd volumes of his collected works. In the next chapter, we shall take a look at Euler's brilliant contributions to a very different branch of mathematics—the field of number theory.

10

Chapter

A Sampler of Euler's Number Theory

(1736)

The Legacy of Fermat

We have seen Euler's success at evaluating complicated infinite series. This work falls within the mathematical subfield called "analysis," where his discoveries were particularly important and profound. But we would be remiss not to examine a few of his contributions to the theory of numbers, a branch of mathematics in which Euler's name ranks very near the top. We encountered number theory before, in Chapter 3 with Euclid's masterful proof of the infinitude of primes and in Chapter 7 with the tantalizing work of Pierre de Fermat, whose insightful commentaries and conjectures transformed the subject. As noted, Fermat was not forthcoming with the proofs, and in the century separating Fermat and Euler, little progress had been made toward verifying Fermat's assertions. This absence of progress may be explained in part by the excitement of the newly discovered calculus, which monopolized mathematical inquiry at the end of the seventeenth century, in part by the perceived lack of applications of number theory to any real-world phenomena, and in part by the humbling fact that Fermat's claims were too difficult for many mathematicians to tackle.

Euler's enthusiasm for number theory was nurtured by Christian Goldbach, whose Goldbach conjecture we briefly met in the Epilogue to Chapter 3. Goldbach was fascinated by the subject of numbers, although his zeal far outstripped his talents. He and Euler maintained a steady correspondence, and it was Goldbach who initially brought many of Fermat's unproved statements to Euler's attention. Initially, Euler seemed less than enthusiastic about pursuing the subject, but a combination of his own insatiable curiosity and Goldbach's persistence forced Euler to take a closer look. Before long, he was captivated by number theory in general and by Fermat's list of unproved statements in particular. As observed by modern author and mathematician Andre Weil, ". . . a substantial part of Euler's [number theoretic] work consisted in no more, and no less, than getting proofs for Fermat's statements." Before he was done, Euler's number theory filled four large volumes of his *Opera Omnia*. It has been observed that, had he done nothing else in his scientific career, these four volumes would place him among the greatest mathematicians of history.

For instance, Euler proved one of Fermat's fascinating assertions about those primes that can be written as the sum of two perfect squares. It is clear that, except for 2, all other primes are odd numbers. Of course, when we divide an *odd* number by 4, we must get a remainder of either 1 or 3 (since numbers that are exact multiples of 4, or two more than multiples of 4, are even.) We can express this more succinctly by saying that, if $p > 2$ is prime, then either $p = 4k + 1$ or $p = 4k + 3$ for some whole number k. In 1640, Fermat had asserted that primes of the first type—that is, those that are one more than a multiple of 4—can be written as the sum of two perfect squares in one and only one way, while primes of the form $4k + 3$ cannot be written as the sum of two perfect squares in any fashion whatever.

This is a peculiar theorem. It says, for instance, that a prime like $193 = (4 \times 48) + 1$ can be written as the sum of two squares in a unique way. In this case, it is easily verified that $193 = 144 + 49 = 12^2 + 7^2$ and that there is no other sum of squares totaling 193. On the other hand, the prime $199 = (4 \times 49) + 3$ cannot be written as the sum of two squares at all, which can likewise be checked by listing all the unsuccessful possibilities. We thus have a major split between the two types of (odd) primes in their expressibility as the sum of two squares. It is a property that is in no way expected or intuitively plausible. Yet Euler had proved it by 1747.

We saw another example of Euler's number theoretic genius when his characterization of all even perfect numbers was discussed in the

Epilogue to Chapter 3. Related to this topic was his work with the so-called *amicable* numbers. These are a pair of numbers with the following property: the sum of all proper divisors of the first number exactly equals the second number while the sum of all proper divisors of the second likewise equals the first. Amicable pairs had been of interest as far back as classical times, when they were regarded by some as having a mystical, "extra-mathematical" significance. Even in the present day, they loom large in the pseudoscience of numerology because of their unusual reciprocal property.

The Greeks were aware that the numbers 220 and 284 were amicable. That is, the proper divisors of 220 are 1, 2, 4, 5, 10, 11, 20, 22, 44, 55, and 110, which add up to 284; at the same time, the divisors of 284 are 1, 2, 4, 71, and 142, whose sum is 220. Unfortunately for numerologists, no other pair was known until Fermat demonstrated in 1636 that 17,296 and 18,416 form a second such pair. [Actually, this pair had been discovered by the Arab mathematician al-Banna (1256–1321) over three centuries earlier, but its existence remained unknown in the West when Fermat came upon the scene.] Then in 1638, Descartes, perhaps in an effort to upstage his countryman, proudly announced his discovery of a third pair, 9,363,584 and 9,437,056.

So matters stood for a century until Euler turned his attention to the problem. Between 1747 and 1750, he found the pair 122,265 and 139,815 as well as 57 other amicable pairs, thereby single-handedly increasing the world's known supply by nearly 2000 percent! What happened was that Euler had found a recipe for generating such numbers, and generate them he did.

One of the most important of all of Fermat's assertions appeared in another letter of 1640. There he stated that, if a is any whole number and p is a prime that is not a factor of a, then *p must be* a factor of the number $a^{p-1} - 1$. As was his irksome custom, Fermat announced that he had found a proof of this strange fact, but did not include it in the letter. Instead he told his correspondent, "I would send you the demonstration if I did not fear it being too long."

This result has since come to be known as the "little Fermat theorem." For the prime $p = 5$ and the number $a = 8$, the theorem asserted that 5 divides evenly into $8^4 - 1 = 4096 - 1 = 4095$; obviously, this is true. Similarly for the prime $p = 7$ and the number $a = 17$, his result claimed that 7 divides evenly into $17^6 - 1 = 24{,}37{,}569 - 1 = 24{,}137{,}568$; this fact is far less obvious but equally true.

How Fermat went about proving this we can only conjecture. A complete proof had to await Euler in 1736. We shall examine this argument in a moment. But first we should assemble the number-theoretic ingredients that Euler needed in order to cook up his proof:

(A) If p is a prime that divides evenly into the product $a \times b \times c \times \cdots \times d$, then p must divide evenly into (at least) one of the factors $a, b, c, \ldots, d$. In everyday language, this says that if a prime divides a product, it must divide one of the factors. As we noted in Chapter 3, Euclid had proved this two millennia before Euler as Proposition VII.30 of the *Elements*.

(B) If p is a prime and a is any whole number, then the expression

$$a^{p-1} + \frac{p-1}{2 \times 1} a^{p-2} + \frac{(p-1)(p-2)}{3 \times 2 \times 1} a^{p-3} + \cdots + a$$

is a whole number as well.

We shall not prove this statement but will instead investigate its truth for a particular example or two. For instance, if $a = 13$ and $p = 7$, then we find that

$$13^6 + \frac{6}{2 \times 1} 13^5 + \frac{6 \times 5}{3 \times 2 \times 1} 13^4 + \frac{6 \times 5 \times 4}{4 \times 3 \times 2 \times 1} 13^3$$
$$+ \frac{6 \times 5 \times 4 \times 3}{5 \times 4 \times 3 \times 2 \times 1} 13^2 + \frac{6 \times 5 \times 4 \times 3 \times 2}{6 \times 5 \times 4 \times 3 \times 2 \times 1} 13 = 4826809$$
$$+ 1113879 + 142805 + 10985 + 507 + 13 = 6094998$$

which indeed is a whole number. What happened here was that all of the apparent fractions in the initial expression canceled out, and we were left with the sum of integers. Of course, it is not obvious that such a cancellation will always occur. Indeed, if we use a nonprime in place of p, we can run into trouble. For instance, with $a = 13$ and $p = 4$, we get

$$13^3 + \frac{3}{2 \times 1} 13^2 + \frac{3 \times 2}{3 \times 2 \times 1} 13 = 2197 + 253.5 + 13 = 2463.5$$

which is certainly not an integer. It is the primality of p that keeps the expression integer-valued.

The only other mathematical weapon that Euler needed was the binomial theorem applied to $(a + 1)^p$. Fortunately, he had read his Newton, so this was in his well-stocked arsenal. We shall approach his argument in a series of four steps, each leading directly to the next and ending with the little Fermat theorem:

THEOREM If p is prime and a is any whole number, then $(a + 1)^p - (a^p + 1)$ is evenly divisible by p.

PROOF Expanding the first expression by the binomial theorem yields

$$(a+1)^p = \left[a^p + pa^{p-1} + \frac{p(p-1)}{2 \times 1}a^{p-2} + \frac{p(p-1)(p-2)}{3 \times 2 \times 1}a^{p-3} + \cdots + pa + 1 \right]$$

We substitute this expansion into the expression $(a+1)^p - (a^p + 1)$, combine terms, and factor out p to get

$$\begin{aligned}(a+1)^p - (a^p+1) &= \left[a^p + pa^{p-1} + \frac{p(p-1)}{2 \times 1}a^{p-2} + \frac{p(p-1)(p-2)}{3 \times 2 \times 1}a^{p-3} + \cdots + pa + 1 \right] - (a^p+1) \\ &= pa^{p-1} + \frac{p(p-1)}{2 \times 1}a^{p-2} + \frac{p(p-1)(p-2)}{3 \times 2 \times 1}a^{p-3} + \cdots + pa \\ &= p\left[a^{p-1} + \frac{p-1}{2 \times 1}a^{p-2} + \frac{(p-1)(p-2)}{3 \times 2 \times 1}a^{p-3} + \cdots + a \right]\end{aligned}$$

But the term in the square brackets is a whole number, by observation (B) above. We have thus demonstrated that $(a+1)^p - (a^p + 1)$ can be factored as the prime p times a whole number. In other words, p divides evenly into $(a+1)^p - (a^p + 1)$, as claimed.

Q.E.D.

This brings us to the second theorem in the sequence.

THEOREM If p is prime and if $a^p - a$ is evenly divisible by p, then so is $(a+1)^p - (a+1)$.

PROOF The previous result guarantees that p divides evenly into $(a+1)^p - (a^p + 1)$, and we have assumed that p also divides evenly into $a^p - a$. Thus, p clearly divides evenly into the sum of these:

$$\begin{aligned}[(a+1)^p - (a^p+1)] + [a^p - a] &= (a+1)^p - a^p - 1 + a^p - a \\ &= (a+1)^p - (a+1)\end{aligned}$$

which is what we were to prove.

Q.E.D.

The previous result provided Euler with the key to proving the little Fermat theorem by a process called "mathematical induction." Induction is a technique of proof ideally suited to propositions involving the whole numbers, for it exploits the "stairstep" nature of these numbers, where one follows immediately after its predecessor. Inductive proofs are much like climbing a (very tall) ladder. Our initial job is to step onto the ladder's first rung. We then must be able to go from the first rung to the second. That accomplished, we need to climb from the second to the third, then from the third to the fourth. If we have mastered the process of climbing from any rung to the next, the ladder is ours! We are assured that there is no rung beyond our reach. So it was with Euler's inductive proof:

THEOREM If p is prime and a is any whole number, then p divides evenly into $a^p - a$.

PROOF Since this was a proposition about *all* whole numbers, Euler began by verifying it for the first whole number, $a = 1$. But this case is simple since $a^p - a = 1^p - 1 = 1 - 1 = 0$, and p certainly divides evenly into 0 (in fact, every positive integer divides evenly into 0). This put him onto the ladder.

Now apply the preceding theorem with $a = 1$—that is, having just established that p is a factor of $1^p - 1$, Euler could conclude that p was likewise a factor of

$$(1 + 1)^p - (1 + 1) = 2^p - 2$$

In other words, his result holds for $a = 2$. Cycling back through the previous proposition, we find that this implies that p divides evenly into

$$(2 + 1)^p - (2 + 1) = 3^p - 3$$

Repeat the process to see that p divides evenly into $4^p - 4$, and $5^p - 5$, and so on. Like climbing from one rung to the next, Euler could proceed up the ladder of whole numbers, assured that, for any whole number a, p is a factor of $a^p - a$.

Q.E.D.

Finally, Euler was ready to give a proof of the little Fermat theorem. Having done the spadework above, he had an easy time of it:

LITTLE FERMAT THEOREM If p is prime and a is a whole number which does not have p as a factor, then p divides evenly into $a^{p-1} - 1$.

PROOF We have just shown that p divides evenly into

$$a^p - a = a[a^{p-1} - 1]$$

Since p is a prime, (A) above implies that p must divide evenly into either a or $a^{p-1} - 1$ (or both). But we have assumed that p does not go evenly into the former, and we are forced to conclude that p does divide evenly into the latter. That is, p divides into $a^{p-1} - 1$. This is the little Fermat theorem.

Q.E.D.

Euler's argument was a gem. He needed only relatively simple concepts; he included an inductive portion, so typical of proofs about whole numbers; he used a result as old as Euclid and another as fresh as the binomial theorem. To these ingredients, he added a liberal dose of his own genius, and out came the first demonstration of Fermat's previously stated, but hitherto unproven, little theorem.

A brief aside is the surprising fact that this proposition has recently been applied to a real-world problem—namely, the design of some highly sophisticated encryption systems for transmitting classified messages. This is not the first and certainly not the last case where an abstract theorem of pure mathematics has proved to have some very down-to-earth uses.

Great Theorem: Euler's Refutation of Fermat's Conjecture

For our purposes in this chapter, the preceding argument is prologue. It was in the context of another Fermat/Euler result that the chapter's great theorem appeared. Not surprisingly, the matter was brought to Euler's attention by his faithful correspondent Goldbach. In a letter of December 1, 1729, Goldbach somewhat innocently asked, "Is Fermat's observation known to you, that all numbers $2^{2^n} + 1$ are primes? He said he could not prove it; nor has anyone else done so to my knowledge."

What Fermat purported to have found was a formula that always generated primes. Clearly he was on target for the first few values of n. That is, if $n = 1$, $2^{2^1} = 2^2 + 1 = 5$ is prime; for $n = 2$, we have $2^{2^2} + 1 = 2^4 + 1 = 17$, a prime; and likewise $n = 3$ and $n = 4$ yield primes $2^8 + 1 = 257$ and $2^{16} + 1 = 65537$. The next number in the sequence, when $n = 5$, is the monster

$$2^{2^5} + 1 = 2^{32} + 1 = 4{,}294{,}967{,}297$$

which Fermat likewise suggested was a prime. Given Fermat's track record, there was no immediate reason to doubt his observation. On the other hand, any attempt to *disprove* Fermat's conjecture would require a mathematician to find a way to factor this 10-digit number into two smaller pieces; such a search could take months, and of course if Fermat were correct about the number's primality, the quest would ultimately prove fruitless. In short, there was every reason to accept Fermat at his word and go on to other business.

But this was not for Leonhard Euler. Instead, he turned his attention to 4,294,967,297, and when the dust settled, Euler had successfully factored it. Fermat had been wrong. Needless to say, Euler did not stumble upon the factors by accident. Like a detective who first eliminates the innocent by-standers from the true suspects in a case, so Euler devised a highly ingenious test that allowed him to eliminate at the outset all but a handful of potential divisors of 4,294,967,297. His extraordinary insight made the task ahead of him immeasurably simpler.

Euler's line of attack began with an even number a (although, if the truth be known, he was thinking specifically of $a = 2$) and a prime p that is not a factor of a. He then wanted to determine the restrictions on this prime p if it *did* divide evenly into $a + 1$, or into $a^2 + 1$, or $a^4 + 1$, or in general into $a^{2^n} + 1$. Given the nature of Fermat's assertion, Euler was particularly interested in the case when $n = 5$. That is, what could he learn about prime factors of $a^{32} + 1$?

It is a perverse twist of fate that the main result Euler used to refute Fermat's conjecture about $2^{2^n} + 1$ was none other than the little Fermat theorem itself. Put another way, Fermat had sown the seeds of his own downfall. Indeed, as we watch Euler reason his way through the great theorem below, we cannot help but admit that, in the right hands, a little Fermat goes a long way.

THEOREM A Suppose a is an even number and p is a prime that is not a factor of a but that does divide evenly into $a + 1$. Then for some whole number k, $p = 2k + 1$.

PROOF This is quite simple. If a is even, then $a + 1$ is odd. Since we assumed that p divides *evenly* into the odd number $a + 1$, p itself must be odd. Hence $p - 1$ is even and so $p - 1 = 2k$ for some whole number k. In other words, $p = 2k + 1$.

Q.E.D.

Consider a specific numerical example. If we start with the even number $a = 20$, then $a + 1 = 21$, and the prime factors of 21—namely, 3 and 7—are indeed both of the form $2k + 1$.

The next step is more challenging:

THEOREM B Suppose a is an even number and p is a prime that is not a factor of a but such that p does divide evenly into $a^2 + 1$. Then for some whole number k, $p = 4k + 1$.

PROOF Since a is even, so is a^2, and by Theorem A we know that any prime factor of $a^2 + 1$—in particular the number p—must be odd. That is, p is one more than a multiple of 2.

But what happens when we divide p by 4? Obviously, any odd number is either one more than a multiple of 4 or three more than a multiple of 4. Symbolically, we can say that p is either of the form $4k + 1$ or of the form $4k + 3$.

Euler wanted to eliminate the latter possibility, and so for the sake of eventual contradiction, he supposed that $p = 4k + 3$ for some whole number k. By hypothesis, p is not a divisor of a, and so the little Fermat theorem implies that p does divide evenly into

$$a^{p-1} - 1 = a^{(4k+3)-1} - 1 = a^{4k+2} - 1.$$

On the other hand, we have assumed that p is a divisor of $a^2 + 1$, and consequently p is also a divisor of the product

$$(a^2 + 1)(a^{4k} - a^{4k-2} + a^{4k-4} - \cdots + a^4 - a^2 + 1).$$

It can be checked algebraically that, upon multiplying out and canceling terms, this complicated product reduces to the relatively simple $a^{4k+2} + 1$.

We have now concluded that p divides evenly into both $a^{4k+2} + 1$ and $a^{4k+2} - 1$. Hence p must be a divisor of the difference

$$(a^{4+2} + 1) - (a^{4k+2} - 1) = 2$$

But this is a glaring contradiction, since the odd prime p cannot divide evenly into 2. The contradiction implies that p does not have the form $4k + 3$, as we assumed at the outset. Since there is only one remaining alternative, we conclude that p must be of the form $4k + 1$ for some whole number k.

Q.E.D.

As before, a few examples are in order. If a = 12, then $a^2 + 1 = 144 + 1 = 145 = 5 \times 29$, and both 5 and 29 are primes of the form $4k + 1$ (that is, one more than a multiple of 4). Alternately, if a = 68, then $a^2 + 1 = 4625 = 5 \times 5 \times 5 \times 37$, and again each of these prime factors is one more than a multiple of 4.

Next, Euler addressed prime factors of the number $a^{2^2} + 1 = a^4 + 1$.

THEOREM C Suppose a is an even number and p is a prime that is not a factor of a but such that p does divide evenly into $a^4 + 1$. Then for some whole number k, $p = 8k + 1$.

PROOF First note that $a^4 + 1 = (a^2)^2 + 1$. Consequently, we can apply Theorem B to deduce that p is one more than a multiple of 4. With this in mind, Euler inquired what would happen if p is divided not by 4 but by 8. At first, we seem to encounter eight possibilities:

$p = 8k$ (i.e., p is a multiple of 8)
$p = 8k + 1$ (i.e., p is one more than a multiple of 8)
$p = 8k + 2$ (i.e., p is two more than a multiple of 8)
$p = 8k + 3$ (i.e., p is three more than a multiple of 8)
$p = 8k + 4$ (i.e., p is four more than a multiple of 8)
$p = 8k + 5$ (i.e., p is five more than a multiple of 8)
$p = 8k + 6$ (i.e., p is six more than a multiple of 8)
$p = 8k + 7$ (i.e., p is seven more than a multiple of 8)

Fortunately—and this was at the heart of Euler's analysis—we can eliminate some of these as possible forms for p. First of all, we know that p must be odd (since it is a divisor of the odd number $a^4 + 1$), and so p cannot take the form $8k$, $8k + 2$, $8k + 4$, or $8k + 6$, all of which are clearly even.

Moreover, $8k + 3 = 4(2k) + 3$ is three more than a multiple of 4, and we know from Theorem B that p cannot take this form. Likewise the number $8k + 7 = 8k + 4 + 3 = 4(2k + 1) + 3$ is also three more than a multiple of 4, and it too is removed from consideration.

So the only possible prime divisors of $a^4 + 1$ are of the form $8k + 1$ or $8k + 5$. But Euler succeeded in eliminating the latter case as follows:

Suppose, for the sake of contradiction, that $p = 8k + 5$ for some whole number k. Then, since p is not a divisor of a, the little Fermat theorem says that p does divide evenly into

$$a^{p-1} - 1 = a^{(8k+5)-1} - 1 = a^{8k+4} - 1$$

On the other hand, since p divides evenly into $a^4 + 1$, it surely divides evenly into

$$(a^4 + 1)(a^{8k} - a^{8k-4} + a^{8k-8} - a^{8k-12} + \cdots + a^8 - a^4 + 1)$$

and this product reduces algebraically to $a^{8k+4} + 1$. But if p is a factor of both $a^{8k+4} + 1$ and $a^{8k+4} - 1$, then p is a factor of their difference

$$(a^{8k+4} + 1) - (a^{8k+4} - 1) = 2$$

and this is a contradiction since p is an odd prime. As a consequence, we see that p cannot have the form $8k + 5$, and so the only possibility for p is, as the theorem asserted, $8k + 1$.

Q.E.D.

Again we consider a quick example. Take $a = 8$ as the even number in question. Then $a^4 + 1 = 4097$, which factors into 17×241, and both factors are one more than a multiple of 8.

From here, Euler established a few more cases in similar fashion, but for our purposes, the pattern should be clear. We can summarize all the previous work as follows. For a an even number and p a prime, then

if p divides evenly into $a + 1$, p is of the form $2k + 1$ (Theorem A)
if p divides evenly into $a^2 + 1$, p is of the form $4k + 1$ (Theorem B)
if p divides evenly into $a^4 + 1$, p is of the form $8k + 1$ (Theorem C)
if p divides evenly into $a^8 + 1$, p is of the form $16k + 1$
if p divides evenly into $a^{16} + 1$, p is of the form $32k + 1$
if p divides evenly into $a^{32} + 1$, p is of the form $64k + 1$

In general, if p divides evenly into $a^{2^n} + 1$, then $p = (2^{n+1})k + 1$ for some whole number k.

At last, we can return to Fermat's conjecture about the primality of $2^{32} + 1$. We return, however, armed with very specific information about the nature of any prime factors this number might have. Rather than groping for factors, Euler could move rather quickly to the heart of the matter.

THEOREM $2^{32} + 1$ is not prime.

PROOF Since $a = 2$ is certainly even, the preceding work tells us that any prime factor of $2^{32} + 1$ must take the form $p = 64k + 1$, where k is a whole number. We thus can check these highly specialized numbers individually to see if they (1) are prime and (2) divide evenly into 4,294,967,297 (Euler did the latter by long division, although the modern reader may want to use a calculator):

if $k = 1$, $64k + 1 = 65$, which is not prime and thus need not be checked
if $k = 2$, $64k + 1 = 129 = 3 \times 43$, again not a prime at all
if $k = 3$, $64k + 1 = 193$, which is a prime but does not divide into $2^{32} + 1$
if $k = 4$, $64k + 1 = 257$, a prime which also fails to divide into $2^{32} + 1$
if $k = 5$, $64k + 1 = 321 = 3 \times 107$, a non-prime which need not be checked
if $k = 6$, $64k + 1 = 385 = 5 \times 7 \times 11$, so we move on
if $k = 7$, $64k + 1 = 449$, a prime which does not divide into $2^{32} + 1$
if $k = 8$, $64k + 1 = 513 = 3 \times 3 \times 3 \times 19$, so go to the next case
if $k = 9$, $64k + 1 = 577$, a prime but not a factor of $2^{32} + 1$

But, when Euler tried $k = 10$, he hit paydirt. In this case we have $p = (64 \times 10) + 1 = 641$, a prime that—lo and behold—divides perfectly into Fermat's number. That is,

$$2^{32} + 1 = 4,294,967,297 = 641 \times 6,700,417$$

Q.E.D.

It is significant that the factor Euler found, 641, was only the fifth number he tried. By carefully eliminating potential divisors of $2^{32} + 1$, he had so depleted the list of suspects that his task became almost trivial. It was a spectacular example of mathematical detective work.

An interesting postscript is based on Euler's previously mentioned theorem that primes of the form $4k + 1$ have a unique decomposition into the sum of two squares. First, we observe that

$$2^{32} + 1 = (2^2)(2^{30}) + 1 = 4(1073741824) + 1$$

and so $2^{32} + 1$ indeed has the form $4k + 1$. But it is straightforward to check numerically that

$$2^{32} + 1 = 4,294,967,297 = 4,294,967,296 + 1 = 65536^2 + 1^2$$

while simultaneously,

$$\begin{aligned} 2^{32} + 1 = 4,294,967,297 &= 418,161,601 + 3,876,805,696 \\ &= 20449^2 + 62264^2 \end{aligned}$$

Here we have decomposed the number $2^{32} + 1$ into the sum of two perfect squares in two *different* ways. By Euler's criterion, this proves that $2^{32} + 1$ cannot possibly be a prime, since primes of the form $4k + 1$ have just one such decomposition. Thus, without finding any explicit factors, we nonetheless have a wonderfully round-about verification that this huge number is composite.

Fermat's assertion that $2^{2^n} + 1$ is prime for all whole numbers n is false for $n = 5$. What is the situation for larger values of n? For $n = 6$, it turns out that

$$2^{2^6} + 1 = 2^{64} + 1 = 18{,}446{,}744{,}073{,}709{,}551{,}617$$

is divisible by the prime $p = 274{,}177$. Not surprisingly, given the pattern Euler had discovered, p is of the form $128k + 1$; that is, $p = (128 \times 2142) + 1$. Fermat was wrong again.

The situation got even worse. By a very sophisticated argument, it was shown in 1905 that the next of Fermat's numbers—$2^{2^7} + 1 = 2^{128} + 1$—is also composite, although the proof did not provide an explicit divisor of this enormous number. Such a factor was not found until 1971, and this factor alone runs to 17 digits.

As of 1988, mathematicians know that $2^{2^8} + 1, 2^{2^9} + 1, \ldots, 2^{2^{21}} + 1$ are all composite. Clearly, Fermat's sweeping conjecture about numbers of the form $2^{2^n} + 1$ has been—so to speak—swept under the rug. Whereas he asserted that *all* of these numbers are prime, no such primes have yet been found for $n \geq 5$. In fact, many mathematicians would now surmise that *none* of these numbers is prime, except of course for Fermat's four primes corresponding to $n = 1, 2, 3$, and 4. This would make his conjecture not just wrong, but very wrong indeed.

With this we shall conclude our short survey of Euler's number theory. As noted, the results on these pages give just the barest hint of Euler's enormous influence in the field. To be sure, he stood on the shoulders of talented predecessors, particularly Fermat. But, by the time Euler was done, he had immeasurably enriched this branch of mathematics and established himself as a number theorist of the highest order.

Epilogue

The year Euler died, Carl Friedrich Gauss turned six years old. Already, the German boy had impressed his elders with his intellectual ability, and in the decades to come he would inherit Euler's mantle as the world's leading mathematician.

Chapter 3 mentioned Gauss's earliest significant achievement, his discovery in 1796 that a regular 17-sided polygon could be constructed with compass and straightedge. This proof caused a sensation in the mathematical world, since no one from classical times forward had the least suspicion that such a construction was possible. We should let young Gauss speak for himself on this point:

> It is well known to every beginner in Geometry that various regular polygons can be constructed geometrically, namely the triangle, pentagon, 15-gon, and those which arise from these by repeatedly doubling the number of sides. One had already got this far in Euclid's time, and it seems that one has persuaded oneself ever since that the domain of elementary geometry could not be extended. . . . It seems to me then to be all the more remarkable that besides the usual polygons there is a collection of others which are constructible geometrically, e.g., the 17-gon.

Gauss, not yet 20 years old, had been able to see more deeply than Euclid, Archimedes, Newton, or anyone else who had ever thought about constructing regular polygons.

But he did more than demonstrate the constructibility of the regular 17-gon, for Gauss determined that a regular polygon with *N* sides was constructible if *N* was a prime number of the form $2^{2^n} + 1$. Of course, as we have just seen, these are precisely Fermat's alleged primes. Somehow, this number theoretic topic turned out to be intimately linked to geometric constructions. As sometimes happens in mathematics, the discoveries and investigations in one branch of the subject—in this case in number theory—played a role in an apparently unrelated branch—geometric constructions of regular polygons. Of course, the key words here are "apparently unrelated." As a matter of fact, Gauss's work showed an undeniable relation indeed.

His discovery, then, not only gave the world the constructibility of the regular 17-gons, since $2^{2^2} + 1 = 17$ is prime, but also of regular $2^{2^3} + 1 = 257$-gons and even the stupendous $2^{2^4} + 1 = 65537$-gons! These constructions, of course, had absolutely no practical utility, but their very existence hinted yet again at the strange, unexpected world lurking beneath the familiar surface of Euclidean geometry. Gauss himself was so proud of this discovery that, even after a lifetime of extraordinary mathematical achievement, he requested that a regular 17-gon be inscribed upon his tombstone. (Unfortunately, it was not.)

Born in 1777 in Brunswick, Carl Friedrich Gauss showed early and unmistakable signs of being an extraordinary youth. As a child of three, he was checking, and occasionally correcting, the books of his father's business, this from a lad who could barely peer over the desk top into the ledger. A famous and charming story is told of Gauss's elementary school training. One of his teachers, apparently eager for a respite from the day's lessons, asked the class to work quietly at their desks and add up the first hundred whole numbers. Surely this would occupy the little tykes for a good long time. Yet the teacher had barely spoken, and the other children had hardly proceeded past "$1 + 2 + 3 + 4 + 5 = 15$" when Carl walked up and placed the answer on the teacher's desk. One imagines that the teacher registered a combination of incredulity and frustration at this unexpected turn of events, but a quick look at Gauss's answer showed it to be perfectly correct. How did he do it?

First of all, it was not magic, nor was it the ability to add a hundred numbers with lightning speed. Rather, even at this young age, Gauss exhibited the penetrating insight that would remain with him for a lifetime. As the story goes, he simply imagined the sum he sought—which we shall denote by S—being written simultaneously in ascending and in descending order:

$$S = 1 + 2 + 3 + 4 + \cdots + 98 + 99 + 100$$
$$S = 100 + 99 + 98 + 97 + \cdots + 3 + 2 + 1$$

Instead of adding these numbers horizontally across the rows, Gauss added them vertically down the columns. In so doing, of course, he got

$$2S = 101 + 101 + 101 + \cdots + 101 + 101$$

since the sum of each column is just 101. But there are a hundred columns. Thus, $2S = 100 \times 101 = 10100$, and so the sum of the first hundred whole numbers is just

$$S = 1 + 2 + 3 + \cdots + 99 + 100 = \frac{10100}{2} = 5050$$

All of this went through Gauss' little head in a flash. It was clear that he was going to make a name for himself.

Accelerating his youthful studies, and under the patronage of the much-impressed Duke of Brunswick, Gauss was in college at 15 and at the prestigious Göttingen University three years later. It was while there that he made the extraordinary discovery about the 17-gons in 1796. This apparently was decisive in turning him to a career in mathematics; he had previously flirted with the idea of becoming a philologist, but the 17-gon convinced him that, perhaps, he was meant to do math.

In 1799, Gauss received his doctorate from the University of Helmstadt for providing the first reasonably complete proof of what is now called the fundamental theorem of algebra. By its name alone we get some sense of the theorem's importance. The proposition deals with the solutions of polynomial equations, obviously a fundamental algebraic topic if ever there was one.

Although versions appeared as early as the seventeenth century, the fundamental theorem of algebra was raised to prominence by the French mathematician Jean d'Alembert (1717–1783), who tried but failed to supply a proof in 1748. He stated the theorem as follows: Any polynomial with real coefficients can be factored into the product of real linear and/or real quadratic factors. For example, the factorization

$$3x^4 + 5x^3 + 10x^2 + 20x - 8 = (3x - 1)(x + 2)(x^2 + 4)$$

illustrates the kind of decomposition d'Alembert had in mind. The real polynomial in question has been broken into simpler pieces: two linear and one quadratic.

Anticipating a bit, we observe that we can further factor the *quadratic* expression, provided we allow ourselves the luxury of complex numbers. We have seen such numbers already in our discussion of the cubic equation, and they figure prominently in the subsequent history of the fundamental theorem of algebra. One can check that, if a, b, and c are real numbers with $a \neq 0$, then

$$ax^2 + bx + c = \left(ax - \frac{-b + \sqrt{b^2 - 4ac}}{2}\right)\left(x - \frac{-b - \sqrt{b^2 - 4ac}}{2a}\right)$$

Here the real quadratic $ax^2 + bx + c$ has been split into two rather unsightly linear factors. (The perceptive reader will see the quadratic formula at work in this factorization.)

Of course, there is no guarantee that these linear factors are composed of *real* numbers, for if $b^2 - 4ac < 0$, we plunge into the realm of imaginaries. In the specific example cited above, for instance, we can further decompose the quadratic term to get the complete factorization:

$$3x^4 + 5x^3 + 10x^2 + 20x - 8$$
$$= (3x - 1)(x + 2)(x - 2\sqrt{-1})(x + 2\sqrt{-1})$$

This is complete in the sense that the real fourth-degree polynomial with which we began has been factored into the product of four *linear* factors, certainly as far as any factorization can hope to proceed. From this vantage point, the fundamental theorem of algebra states that any real polynomial of degree n can be factored into n (perhaps complex) linear factors.

As noted, d'Alembert recognized the importance of such a statement and made a stab at a proof. His stab, unfortunately, was wide of the mark. Perhaps to accord him the honor of trying, this result was long known as "d'Alembert's theorem," in spite of the fact that he came nowhere near proving it. This seems somewhat akin to renaming Moscow after Napoleon simply because he *tried* to reach it.

So matters stood in the middle of the eighteenth century. Mathematicians were divided as to whether the result was true—Goldbach, for instance, doubted its validity—and even those who believed it were unsuccessful at furnishing the proof. Perhaps the closest anyone came was Leonhard Euler in a remarkable paper of 1749.

Euler's "proof" exhibited his characteristic cleverness and ingenuity. It began well enough, as he correctly showed that real quartic or real quintic equations could be factored into real linear or real quadratic components. But as he pursued this elusive theorem toward higher-degree polynomials, he found himself tangled in a thicket of overwhelming complexity. For instance, at one point he had to establish that a certain equation could be solved for an auxiliary variable u that he had previously introduced. Unfortunately, wrote Euler, "The equation which determines the values of the unknown u will necessarily be of the 12870-th degree." He tried to finesse it by a round-about argument, but he left his critics unconvinced. In short, while he made an admirable try, the fundamental theorem of algebra got the better of him. That even Euler suffered setbacks may bring some comfort to those with lesser mathematical abilities (a category that includes virtually everybody).

The fundamental theorem of algebra—the result that establishes the complex numbers as the ultimate realm for factoring polynomials—thus remained in a very precarious state. D'Alembert had not proved it; Euler had given only a partial proof. It was obviously in need of major attention to resolve its validity once and for all.

This brings us back to Gauss's landmark dissertation with the long and descriptive title, "A New Proof of the Theorem That Every Integral Rational Algebraic Function [that is, every polynomial with real coefficients] Can Be Decomposed into Real Factors of the First or Second Degree." He began with a critical review of previous attacks upon this theorem. When addressing Euler's attempted proof, Gauss observed its shortcomings as lacking "the clarity which is required in mathematics." This clarity he tried to provide, not only in his dissertation but also in alternate proofs of the result he published in 1814, 1816, and 1848.

Today, this crucial theorem is viewed in somewhat more generality than in the early nineteenth century. We now transfer it entirely into the realm of complex numbers in this sense: the polynomial with which we begin no longer is required to have *real* coefficients. In general, we consider nth-degree polynomials having real or complex coefficients, such as

$$z^7 + (6\sqrt{-1})z^6 - (2 + \sqrt{-1})z^2 + 19$$

In spite of the apparent increase in sophistication introduced by this modification, the fundamental theorem guarantees that even polynomials of this type can be factored into the product of n linear terms having, of course, complex coefficients.

Gauss's next major work was in number theory, where he followed in the tradition of Euclid, Fermat, and Euler. In 1801 he published his number-theoretic masterpiece, *Disquisitiones Arithmeticae.* Incidentally, he concluded this book with an extended discussion of constructing regular polygons—a discussion that hinged, quite unexpectedly, upon complex numbers—and the relationship of this construction to number theory. Throughout his life, this subject was always especially dear to Gauss, who once asserted that "Mathematics is the queen of the sciences, and the theory of numbers is the queen of mathematics."

Barely 30 years old, already having made landmark discoveries in geometry, algebra, and number theory, Carl Friedrich Gauss was appointed director of the Observatory at Göttingen. He held this position for the rest of his life. The job required him to consider the applications of mathematics to the real world, a vastly different side of the subject from his beloved numbers. Yet here again he excelled. He was instrumental in determining the orbit of the asteroid Ceres; he carefully mapped the earth's magnetic field; along with Wilhelm Weber, Gauss was an early student of magnetism, and physicists today call a unit of magnetic flux a "gauss" in his honor; Weber and Gauss likewise collaborated on the invention of an electric telegraph some years before similar work on a more ambitious scale established the reputation of Samuel F.B. Morse. Gauss's successes in pure mathematics were matched by his achievements in mathematical applications. Like Newton, he managed to succeed brilliantly in both spheres.

One can draw other parallels between Gauss and Sir Isaac, and these fall as much in the domain of the psychologist as the mathematician. Both individuals were known as being rather icy and distant personalities, content to work in relative seclusion. Neither enjoyed teaching very much, although Gauss did direct the doctoral researches of some of the finest mathematicians the nineteenth century was to produce.

In addition, both men avoided the specter of academic controversy. Recall that, as a young man, Newton seemed more willing to be boiled in oil than to undergo the torment of subjecting his work to public scrutiny. Gauss likewise had qualms about clashing with the prevailing scientific opinion, as was most evident in his discovery of non-Euclidean geometry. In the Epilogue to Chapter 2, we noted his concern about the "howl of the Boeotians" if he made his revolutionary views on this subject known. By the early 1800s, Gauss had established himself as the world's leading mathematician. As such, he seemed particularly conscious of the impact of his ideas and the intense scrutiny to which they would be subjected. To come forth with a brilliant proof of the fundamental theorem of algebra was one thing, but to tell the world that triangles may have fewer than 180° was something else again. Gauss simply refused to take such a stand. Like Newton, he folded up his wonderful discoveries and consigned them to the recesses of his desk.

One side of this rigid, conservative man—a rather unexpected side at that—should not be overlooked. It concerns Gauss's encouragement of Sophie Germain (1776–1831), a woman who had overcome a series of obstacles to rise to prominence in the mathematical world of the early nineteenth century. Her story illustrates, in no uncertain terms, the societal attitude that the discipline of mathematics was no place for a woman.

As a child, Germain had been fascinated by the mathematical works she found in her father's library. She was especially intrigued by Plutarch's description of the death of Archimedes, for whom mathematics was more vital than life itself. When she expressed an interest in studying the subject more formally, her parents responded in horror. Forbidden to explore mathematics, Sophie Germain was forced to smuggle books into her room and read them by the candlelight. Her family, discovering these clandestine activities, removed her candles and, for good measure, removed her clothes as well in an attempt to discourage these nocturnal wanderings in a cold and dark room. It is a testament to Germain's love of mathematics, and perhaps to her physical endurance, that not even these extreme measures could keep her down.

As she mastered ever more of the mathematics of her day, Germain was ready to move on to advanced topics. The very idea that she would attend class at college or university seemed preposterous, so she was forced to eavesdrop outside the classroom door, picking up what information she could and borrowing the lecture notes of sympathetic male colleagues within. Few people have ever had a rockier route into the world of higher mathematics.

And yet, Sophie Germain made it. In 1816, her work was impressive enough that she won a prize from the French Academy for her penetrating analysis of the nature of vibrations in elastic plates. In the process, she had disguised her identity with the pseudonym Antoine LeBlanc so as not to give away her unpardonable sin of being a woman. With this pen-name she also began writing to the world's foremost mathematician.

From the outset, Gauss was impressed with the talent of his French correspondent. LeBlanc had obviously read the *Disquisitiones Arithmeticae* with care and offered certain generalizations and extensions of results contained therein. Then in 1807, the majestic Carl Friedrich Gauss learned the true identity of Sophie Germain. Obviously concerned by the impact of this news, she wrote Gauss what sounded a good deal like a confession:

> . . . I have previously taken the name of M. LeBlanc in communicating to you those notes that, no doubt, do not deserve the indulgence with which you have responded. . . . I hope that the information that I have today confided to you will not deprive me of the honour you have accorded me under a borrowed name, and that you will devote a few minutes to write me news of yourself.

It may come as a surprise that Gauss answered with charity and understanding. He admitted to an "astonishment" at seeing his M. LeBlanc "metamorphosed" into Sophie Germain, then went on to reveal a deeper insight into the inequities of the mathematical establishment:

> The taste for the abstract sciences in general and, above all, for the mysteries of numbers, is very rare: this is not surprising, since the charms of this sublime science in all their beauty reveal themselves only to those who have the courage to fathom them. But when a woman, because of her sex, our customs, and prejudices, encounters infinitely more obstacles than men in familiarizing herself with their knotty problems, yet overcomes these fetters and penetrates that which is most hidden, she doubtless has the most noble courage, extraordinary talent, and superior genius.

Continuing in this vein, Gauss heaped praise upon her mathematical works which "have given me a thousand pleasures" But then he continued, "I ask you to take it as a proof of my attention if I dare add a remark to your last letter," and proceeded to point out a mistake in her reasoning. Sophie Germain's mathematics could give Gauss untold pleasures, but the letter left little doubt as to who, in Gauss' mind, was the master.

We should note that Sophie Germain had a productive career even with her identity revealed. In 1831, at the urging of Gauss himself, she was to have been awarded an honorary doctorate from Göttingen. This would have been a brilliant honor for a woman in early-nineteenth-century Germany. Unfortunately, her death prevented this honor from being bestowed.

And what of Carl Friedrich Gauss? He lived to the age of 78, and death came, fittingly, at the Göttingen Observatory he had directed for almost half a century. By the end of his life, his reputation had assumed

Carl Friedrich Gauss (photograph courtesy of The Ohio State University Libraries)

almost mythical proportions, and a reference to the Prince of Mathematicians meant Gauss and no other.

But he himself had a different motto, one that aptly characterized his life and work: *Pauca sed matura* (Few but ripe). Gauss was a man who published relatively little in his lifetime. The unpublished research found among his papers would have made the reputation of a dozen mathematicians. Yet, always mindful of the great expectations his work raised, he chose not to publish a result until it was as perfect and flawless as could be. Gauss did not write as much as Euler, but when he wrote, the mathematical world was advised to take notice. The fruits he left behind—from the regular 17-gon, to the *Disquisitiones,* to the magnificent fundamental theorem of algebra—were as ripe as any that mathematics is likely to see.

11

CHAPTER

The Non-Denumerability of the Continuum

(1874)

Mathematics of the Nineteenth Century

In a strange way, different centuries bring their own different emphases, their own different directions to the flow of mathematical thought. The eighteenth century clearly was the "century of Euler," for he had no rival in dominating the scholarly landscape and providing a legacy for future generations. The nineteenth century, by contrast, had no single preeminent mathematician, although it was blessed with an array of individuals who pushed the frontiers of the subject in dramatic and unexpected directions.

If the 1800s did not belong to a single mathematician, they did have a few overriding themes. It was the century of abstraction and generalization, of a deeper analysis of the logical foundations of mathematics that underlay the wonderful theories of Newton and Leibniz and Euler. Mathematics charted a course ever more independent of the demands, and the limitations, of "physical reality" that had always tied the subject, in some significant way, to the natural sciences.

This drift away from the constraints of the real world may have had as its declaration of independence the emergence of non-Euclidean geometry in the first third of the nineteenth century. In the Epilogue to Chapter 2, we encountered the "strange, new universe" that emerged when Euclid's parallel postulate was jettisoned and replaced by an alternative statement. Suddenly, more than one parallel could be drawn through a point not on a line, similar triangles became congruent, and triangles no longer contained 180°. Yet for all of these seemingly paradoxical features, no one could find a *logical* contradiction in the non-Euclidean world.

When Eugenio Beltrami established that non-Euclidean geometry was as logically consistent as Euclid's own, a bridge of sorts had been crossed. Imagine, for instance, that Mathematician A had spent a career immersed in the study of Euclidean geometry while Mathematician B worked exclusively with the non-Euclidean variety. Both persons would have been involved in tasks of equivalent *logical* validity. Yet the "real world" could not simultaneously be Euclidean and non-Euclidean; one of our mathematicians must have devoted a lifetime of effort to exploring a system that was not "real." Had he or she squandered a career?

Increasingly in the nineteenth century, mathematicians came to feel that the answer was "No." Of course, either the physical universe was Euclidean or it was not, but the resolution of this problem should be left to the physicists. It was an empirical matter, one to be addressed by experimentation and close observation. But it was irrelevant to the logical development of these competing systems. To a mathematician engrossed in the strange and beautiful theorems of non-Euclidean geometry, the beauty was enough. There was no need for the physicist to tell the mathematician which geometry was "real." In the realm of logic, both were.

Thus, the foundational question in geometry had a liberating effect, freeing mathematics from exclusive dependence on the phenomena of the laboratory. In this sense, we observe an interesting parallel to the simultaneous liberation of art from a similar dependence on reality. As the nineteenth century began, the painter's canvas was what it had long been—a "window" through which the viewer looked to see an event or person of interest. The painter, of course, was free to set the mood, to choose the colors and determine the lighting, to emphasize some details while de-emphasizing others; yet ultimately the canvas would serve as a viewing screen through which to examine an instant frozen in time.

During the second half of the century, this attitude changed markedly. Under the influence of such great artists as Paul Cézanne, Paul Gauguin, and Vincent Van Gogh, the canvas took on a life of its own. Painters could regard it as a two-dimensional field on which to ply their painterly skills. Cézanne, for instance, felt free to distort the wonderful pears and apples of his still-lives to enhance the overall effect. When he criticized the great impressionist Claude Monet as being only "an eye," Cézanne was stating that painters henceforth could regard their art as doing more than simply recording what the eye saw.

Painting, in short, was declaring its independence from visual reality, even as mathematics was exhibiting its freedom from the physical world. The parallel is an interesting one, and certainly the philosophical thrust evident in the mathematics of Gauss, Bolyai, and Lobachevski—as well in the painting of Cézanne, Gauguin, and Van Gogh—has been long-lasting and profound. Their visions are still very much with us today.

It must be noted, of course, that there is not a uniform approval of the direction which these developments have taken. Any casual visitor to an art museum in the late twentieth century is sure to hear disparaging remarks about the state of the visual arts, about huge canvases smeared with meaningless blobs of paint, about controversial and sometimes very expensive works that make no pretense whatever of reflecting a real scene. Patrons are known to grumble that modern artists have carried their liberation too far. They long for a familiar portrait or a comfortable landscape.

In this sense, too, art and mathematics have run parallel courses. There are voices in the modern mathematical community that bemoan the state of mathematics today. While relishing the intellectual freedom bequeathed by the non-Euclidean revolution, mathematicians of the twentieth century carried their subject farther and farther from a contact with the real world, until their logical constructs became so abstract and arcane as to be unrecognizable by a physicist or engineer. To many, this trend has transformed mathematics into little more than a pointless exercise in chasing tiny symbols across the page. One of the most vocal critics of this trend is mathematics historian Morris Kline, who wrote:

> Having formulated the abstract theories, mathematicians turned away from the original concrete fields and concentrated on the abstract structures. Through the introduction of hundreds of subordinate concepts, the subject has mushroomed into a welter of smaller developments that have little relation to each other or to the original concrete fields.

Kline is suggesting that mathematics, in exploiting its hard-won freedom from physics, has ventured too far from its roots and thus become a sterile, self-indulgent, inwardly directed discipline. His is a harsh criticism that the mathematical community must surely consider seriously.

In response there can be made an intriguing argument that mathematical theories, no matter how seemingly abstract, often have surprising applications to very solid, real-world phenomena. Even the non-Euclidean revolution, the subject that did so much to sever the bond between mathematics and reality, has found its way into modern physics books, for the relativistic theories of today's cosmologies rely heavily on a non-Euclidean model of the universe. Such a reliance was certainly not foreseen by the nineteenth-century mathematicians who investigated the subject for its own sake, yet it now forms a part of applied mathematics necessary for inclusion in the physicist's tool-kit. Abstract mathematics can pop up in the strangest places.

The debate continues. Eventually, historians may look at today's mathematics as having ventured too far from its ties to the real world. But it is inconceivable that mathematics will ever assume a role entirely subservient to the needs of the other sciences. Mathematical freedom will forever be the legacy of the nineteenth century.

While these issues arose from the creation of non-Euclidean geometry, another major battle was being fought over the logical foundations of the calculus. We recall that this subject had been formalized and codified by Newton and Leibniz in the late seventeenth century and then exploited by Leonhard Euler in the eighteenth. Yet these pioneers, and their talented mathematical contemporaries, had not paid adequate attention to the underpinnings of calculus. Like a skater gliding over thin ice, these mathematicians were sailing along gloriously, yet at any moment a crack in the foundations could spell disaster.

That there was a problem had been clear for a very long time. It lay in the use of "infinitely large" and "infinitely small" quantities so very essential to Newton's fluxions and Leibniz's calculus. One of the key ideas in all of calculus is that of "limit." In some form or other, both differential and integral calculus—not to mention questions of series convergence and continuity of functions—rest upon this notion. The term "limit" is suggestive, with strong intuitive ties, and we regularly talk about the limit of our patience or the limit of our endurance. Yet when we try to make this idea logically precise, difficulties instantly arise.

Newton gave it a try. His concept of a fluxion required him to look at the ratio of two quantities and then to determine what happened to this ratio as both quantities simultaneously moved toward zero. In modern terminology, he was describing the *limit* of the ratio of these quantities, although he used the somewhat more colorful term "ultimate ratio." To Newton, by the ultimate ratio of vanishing quantities

> . . . is to be understood the ratio of the quantities, not before they vanish, nor after, but that with which they vanish.

Of course, this was little help as a mathematical definition. We may agree with Newton that the limit should not be tied to a ratio *before* they vanish, but what on earth did he mean by the ratio *after* they vanish? Newton, it seems, wanted to consider the ratio at the precise instant when both numerator and denominator became zero. Yet at that instant, the fraction 0/0 has no meaning. The situation was a logical morass.

What did Leibniz have to contribute to the subject? He too needed to address the behavior of limits, and he tended to approach the matter through a discussion of "infinitely small quantities." By this he meant quantities that, while not zero, cannot be decreased further. Like the atoms of chemistry, his infinitely small quantities were the indivisible building blocks of mathematics, the next closest thing to zero. The philosophical problems with such an idea clearly troubled Leibniz, and he had to resort to such unintelligible statements as:

> It will be sufficient if, when we speak of . . . infinitely small quantities (i.e., the very least of those within our knowledge), it is understood that we mean quantities that are . . . indefinitely small. . . . If anyone wishes to understand these [the infinitely small] as the ultimate things . . . , it can be done . . . , ay even though he think that such things are utterly impossible; it will be sufficient simply to make use of them as a tool that has advantages for the purpose of calculation, just as the algebraists retain imaginary roots with great profit.

Besides noting Leibniz's bias against complex numbers, we can find in this statement an incredible amount of mathematical gibberish. The imprecision of ideas—particularly the very ideas that underlay his calculus—obviously caused Leibniz a great deal of anxiety.

Already uneasy over the foundations of their subject, mathematicians got a solid dose of ridicule from a clergyman, Bishop George Berkeley (1685–1753). Bishop Berkeley, in his caustic essay *The Analyst, or a Discourse addressed to an Infidel Mathematician,* derided those mathematicians who were ever ready to criticize theology as being based upon unsubstantiated faith, yet who embraced the calculus in spite of its foundational weaknesses. Berkeley could not resist letting them have it:

> All these points [of mathematics], I say, are supposed and believed by certain rigorous exactors of evidence in religion, men who pretend to believe no further than they can see. . . But he who can digest a second or third fluxion, a second or third differential, need not, methinks, be squeamish about any point in divinity.

As if that were not devastating enough, Berkeley added the wonderfully barbed comment:

> And what are these fluxions? The velocities of evanescent increments. And what are these same evanescent increments? They are neither finite quantities, nor quantities infinitely small, nor yet nothing. May we not call them the ghosts of departed quantities . . . ?

Sadly, the foundations of the calculus had come to this—to "ghosts of departed quantities." One imagines hundreds of mathematicians squirming restlessly under this sarcastic phrase.

Gradually the mathematical community had to address this vexing problem. Throughout much of the eighteenth century, they had simply been having too much success—and too much fun—in exploiting the calculus to stop and examine its underlying principles. But growing internal concerns, along with Berkeley's external sniping, left them little choice. The matter had to be resolved.

Thus we find a string of gifted mathematicians working on the foundational questions. The process of refining the idea of "limit" was an excruciating one, for the concept is inherently quite deep, requiring a precision of thought and an appreciation of the nature of the real number system that is by no means easy to come by. Gradually, though, mathematicians chipped away at this idea. By 1821, the Frenchman Augustin-Louis Cauchy (1789–1857) had proposed this definition:

> When the values successively attributed to a particular variable approach indefinitely a fixed value, so as to end by differing from it by as little as one wishes, this latter is called the *limit* of all the others.

Note that Cauchy's definition avoided such imprecise terms as "infinitely small." He did not get himself into the bind of determining what happened at the precise instant the variable reaches the limit. There are no ghosts of departed quantities here. Instead, he simply said that a fixed value is the limit of a variable if we can make the variable differ from the limit by as little as we wish. This so-called "limit avoidance" definition of Cauchy removed the philosophical barriers as to what happened at that moment of reaching the limit. To Cauchy, the issue was irrelevant; all that mattered was that we could get as close to the limit as we wanted.

What made Cauchy's definition so influential was that he went on to use it in proving the major theorems of calculus. With calculus predicated on this definition of limit, mathematicians had come a long way toward addressing the concerns of Bishop Berkeley. Yet even Cauchy's statement needed some fine-tuning. For one thing, it talked about the "approach" of one variable to a limit, and this conjured up vague ideas about motion; if we must rely on intuitive ideas about points *moving around* and *approaching* one another, are we any better off than we were in just relying on the intuitive idea of "limit" itself? Secondly, Cauchy's use of the term "indefinitely" seems a bit—well—indefinite; clarification of the term is needed. Finally, his definition was simply too wordy; there was a need to replace words and phrases with clearly defined and utterly unambiguous symbols.

And so, the last word in shoring up the foundations of the calculus—a process known by the tongue-twisting phrase "the arithmetization of the calculus"—was given by the German mathematician Karl Weierstrass (1815–1897) and his band of disciples. For the school of Weierstrass, to say that "L is the *limit* of the function $f(x)$ as x approaches a" meant precisely:

> for any $\epsilon > 0$, there exists a $\delta > 0$ so that, if $0 < |x - a| < \delta$, then $|f(x) - L| < \epsilon$.

One need not understand this definition in its entirety to recognize that it differed significantly from Cauchy's. It was almost totally symbolic and nowhere required that any quantity move toward any other quantity. It was, in short, a *static* definition of the limit. Further, it was a far cry from the vague, almost amusing statements of Newton and Leibniz quoted earlier. Weierstrass's severely logical definition lacked some of the spice and charm of his predecessors', but it was mathematically solid. Upon this foundation, Weierstrass built the edifice of calculus that remains to the present day.

Cantor and the Challenge of the Infinite

And yet, as so often happens in science, the resolution of one question merely opened the door to another. As mathematicians examined the calculus from this far more rigorous viewpoint, as they relied less and less upon intuitive concepts and more and more upon the ϵs and δs of a Weierstrassian mathematics, they made some highly peculiar and unsettling discoveries.

For instance, consider the distinction between the rational and the irrational numbers. The former are all of the fractions, the numbers that can be represented as ratios of integers. When converted to decimals, rational numbers are easily spotted: their decimal expansions either terminate abruptly (for example, $3/8 = .375$) or exhibit a block of digits that repeat over and over again forever (for example, $3/11 = .27272727\ldots$). On the other hand, the irrationals are those real numbers which cannot be expressed as fractions. Well-known examples of irrationals are $\sqrt{2}$ or π. For irrationals, the decimal expansion neither terminates nor features a repeating block; instead, their decimals proceed endlessly, without a repetitive pattern.

It is possible to show that both the rationals and the irrationals are densely distributed along the number line in the following sense: Between any two rational numbers, there lie infinitely many irrationals and, conversely, between any two irrationals are to be found infinitely many rationals. Consequently, it is easy to conclude that the real numbers must be evenly divided between the two enormous, and roughly equivalent, families of rationals and irrationals.

As the nineteenth century progressed, mathematical discoveries came to light indicating, to the contrary, that these two classes of numbers did not carry equal weight. The discoveries often required very technical, very subtle reasoning. For instance, a function was described that was continuous (intuitively, unbroken) at each irrational point and discontinuous (broken) at each rational point; however, it was also proved that no function exists that is continuous at each rational point and discontinuous at each irrational point. Here was a striking indicator that there was not a symmetry or balance between the set of rationals and the set of irrationals. It showed that, in some fundamental sense, the rationals and irrationals were not interchangeable collections, but to the mathematicians of the day, it was unclear exactly what was going on.

Considerations of this sort, reaching deep into the very nature of the real number system, were the impetus for the theorems we shall consider in this chapter. Cauchy, Weierstrass, and their colleagues had successfully placed the foundations of the calculus upon the more basic notion of "limits," but mathematicians were coming to realize that some of the most important, and fundamental, questions of the calculus rested upon profound properties of *sets.* The man who would tackle this problem, and in the process single-handedly create a marvelous set theory, was the sometimes maligned, sometimes paranoid genius Georg Ferdinand Ludwig Philip Cantor.

Cantor was born in Russia in 1845, but his family moved to Germany when he was 12. Religion was an important component of the Cantor household. The elder Cantor had converted from Judaism to Protestantism, whereas his wife was born a Roman Catholic. With such an eclectic mix of religious perspectives, it is no surprise that young Georg developed a lifelong interest in theological matters. Some of these, particularly those related to the nature of the infinite, would figure in Cantor's adult mathematics.

In addition, Cantor's family showed a marked artistic streak. Music was highly revered, and Cantor had relatives who played in major symphonies. Georg himself was a respectable draftsman, and he left behind some pencil-and-paper drawing showing definite talent. In short, Cantor exhibited what might be called an "artistic" nature.

This sensitive young man excelled in mathematics and in 1867 completed his doctorate at the University of Berlin, where he had studied with Weierstrass and had thoroughly absorbed the rigorous approach to calculus noted above. Cantor's research into the finer points of mathematical analysis led him more and more to consider the intrinsic differences among various sets of numbers. In particular, he realized the importance of devising a means for comparing the sizes of sets.

On a superficial level, comparing sets for size is a trivial consequence of the ability to count. If asked, "Do you have as many fingers on your right hand as on your left?" one could simply count the fingers on each hand separately and, determining that there were five on each, answer in the affirmative. It seems obvious that the primitive technique of "counting" is prerequisite to the more sophisticated notion of "same size" or "equinumerosity." With a deceptively innocent observation, Georg Cantor turned this truism upside down.

To see how he did this, imagine that we live in a culture whose mathematical knowledge is so limited that people can count only as high as "three." In this case, we could not compare left- and right-hand fingers by *counting* them, since our number system does not carry us all the way to "five." Without the ability to count, is the determination of equinumerosity beyond our reach? Not at all. We could answer the question, not by trying to count fingers, but merely by placing our fingers together, left thumb against right thumb, left index against right index, and so on. This would exhibit a perfect one-to-one correspondence, and again we would answer, "Yes, we have as many left-hand fingers as right-hand ones."

Alternately, imagine an audience filtering into a large auditorium. To answer the question of whether there are as many spectators as seats, we could go through the tedious process of counting both audience and chairs and then compare our final counts. But instead, we could simply ask all in attendance to sit down. If each person had a seat and each seat had a person, the answer is "yes," since the very process of *sitting* has exhibited a perfect one-to-one correspondence.

These examples illustrate the critical fact that we need not be able to count objects in sets in order to determine whether or not the sets are equinumerous. On the contrary, the notion of being equally numerous, seen in the light of one-to-one correspondences, becomes the more primitive, fundamental concept; counting, by contrast, is the more sophisticated, advanced one.

Georg Cantor exploited this idea in the following definition:

> Two sets *M* and *N* are *equivalent* . . . if it is possible to put them, by some law, in such a relation to one another that to every element of each one of them corresponds one and only one element of the other.

Modern mathematicians often say sets M and N have the "same power" or "same cardinality" if they meet Cantor's definition of equivalence above. But, terminology aside, the definition is a critical one since it in no way requires that sets M and N be finite; on the contrary, it applies equally well to sets containing infinitely many elements.

With this, Cantor was moving into uncharted territory. Throughout the history of mathematics, the infinite had been regarded with a suspicious if not hostile eye, as a concept better let alone. From the time of the Greeks, up until Cantor's own century, philosophers and mathematicians had recognized only the "potential infinite." That is, they surely would agree that the set of whole numbers is infinite; we will never run out of them, and at any spot among these numbers, we have the ability to move on to the next bigger one. If we think, for instance, of writing down each whole number on a slip of paper and placing it into a (very large) bag, our process would go on endlessly.

But Cantor's predecessors objected to the "completed infinite"—that is, to regarding this process as ever being finished or the bag as ever being full. In the words of Carl Friedrich Gauss:

> . . . I protest above all against the use of an infinite quantity as a *completed* one, which in mathematics is never allowed. The Infinite is only a manner of speaking. . . .

Cantor did not agree. He was perfectly willing to regard the bag of all whole numbers as a self-contained, completed entity to be compared with other infinite sets of objects. Unlike Gauss, he was unwilling to dismiss "infinity" as merely a figure of speech. To Cantor, it was a solid, respectable mathematical concept worthy of the most profound intellectual examination.

And so, armed only with these two fundamental premises—that equal cardinality of sets can be determined via one-to-one matchings and that the completed infinite was a valid concept—Georg Cantor set off on what would become one of the most exhilarating and demanding intellectual journeys of all time. It would lead down strange paths, and portions of the mathematical establishment were only too happy to ridicule his efforts, but he persevered. In the end, Cantor had the talent, and the courage, to confront the infinite face-to-face in an absolutely unprecedented fashion.

To begin, we shall let $\mathbf{N} = \{1, 2, 3, \ldots\}$ be the set of natural numbers and let $\mathbf{E} = \{2, 4, 6, \ldots\}$ be the set of even natural numbers. Note that we are regarding these as completed sets, in spite of their infinite natures. By Cantor's definition, the sets **N** and **E** are easily seen to have the "same cardinality" since we can exhibit a simple one-to-one correspondence between their members:

N:	1	2	3	4	5	...	n	...
	↕	↕	↕	↕	↕		↕	
E:	2	4	6	8	10	...	$2n$	...

This correspondence clearly matches each element of **N** with one and only one even number (namely, its double) and conversely matches each even number with one and only one natural number (namely, its half). To Cantor, it was clear that these infinite sets were of the same size. Of course, this seems paradoxical at the outset, for one imagines that there should be but half as many even numbers as whole numbers. Yet on what basis can we criticize Cantor's deduction? We either reject the notion of the completed infinite, somehow arguing that we cannot even consider the set of natural numbers as a self-contained entity; or we reject the ridiculously simple definition of equal cardinality. If we accept both of these premises, then the conclusion is inescapable: there are no fewer even numbers than natural numbers.

Likewise, if **Z** = {. . . –3, –2, –1, 0, 1, 2, 3, . . .} represents the set of all the integers—positive, negative, and zero—then we can see that **N** and **Z** have the same cardinality from the following one-to-one matching:

N:	1	2	3	4	5	6	7	8	9	...
	↕	↕	↕	↕	↕	↕	↕	↕	↕	
Z:	0	1	−1	2	−2	3	−3	4	−4	...

For this particular correspondence, it can be checked that each natural number n in **N** is matched with its counterpart

$$\frac{1 + (-1)^n(2n - 1)}{4}$$

in **Z**.

At this point, Cantor was ready to make a bold move. He said that any set which could be put into a one-to-one correspondence with **N** was *denumerable* or *countably infinite*. More strikingly, he introduced a new "transfinite" cardinal number to represent the number of items in a denumerable set. The symbol he chose for this transfinite cardinal was $\aleph_0$ (read "aleph-naught"), after the first letter of the Hebrew alphabet.

Here Cantor's study of infinite sets had led him to create a new number, and a new *kind* of number. One imagines many of his contemporaries shaking their heads in pity at the poor fellow who touted such absurd ideas. And yet, think again about the mathematically primitive culture whose people could only count to three. An innovative genius of that culture might, in a burst of insight, extend the number system beyond its known bounds by introducing the new cardinal number five as follows: A set will be said to contain *five* elements if its members can be put into a one-to-one correspondence with the fingers of her right hand.

Such a definition would work perfectly well. It would provide an unambiguous way of determining when a set has five elements (provided her hand remains intact). In this sense, her fingers become the standard reference point for deciding whether sets have five elements. It all seems so very reasonable.

This was precisely what Cantor did, except that he used the system **N** as the benchmark for extending our number system beyond the finite. For him, **N** was the prototypical set having $\aleph_0$ elements. Introducing the symbol $\overline{\mathbf{M}}$ to mean "the cardinality of the set **M**," we see that

$$\overline{\mathbf{N}} = \overline{\mathbf{E}} = \overline{\mathbf{Z}} = \aleph_0$$

What if we next examine the set **Q** of rational numbers? As we noted earlier, the rationals are densely distributed. In this sense, the rationals differ from the integers, which march in lockstep fashion across the line, each one unit away from its predecessor. In fact, between any two integers—say, between 0 and 1—there are *infinitely* many rationals. Thus, anyone would conjecture that the rationals are far more plentiful than the natural numbers.

But Cantor showed that the set of rationals was denumerable; that is, $\overline{\mathbf{Q}} = \aleph_0$. He did it by the simple expedient of providing a one-to-one correspondence between the set of rationals and the set of natural numbers. To see how he generated this correspondence, consider the rational numbers arranged in the following array:

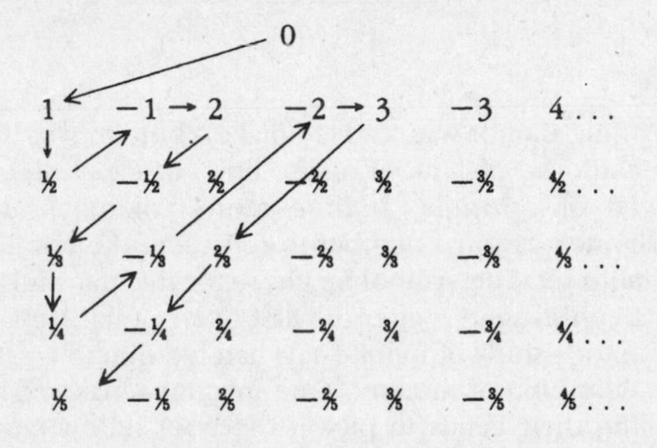

Note that all numbers in the first column have numerator of 1, all in the second column have numerator of −1, and so on, while all numbers in the first row have denominator of 1, all in the second row have denominator 2, etc. In short, given any fraction—for instance $133/191$—we could locate it in the array by going down to the 191st row and running across (counting positives and negatives) to the 265th column. The array, then, contains all of **Q** with redundancy.

But now we exhibit the matching by following the arrows up and down the diagonals, as indicated. This yields the correspondence:

N:	1	2	3	4	5	6	7	8	9	10	11	12	13	14 . . .
	↕	↕	↕	↕	↕	↕	↕	↕	↕	↕	↕	↕	↕	↕
Q:	0	1	$1/2$	-1	2	$-1/2$	$1/3$	$1/4$	$-1/3$	-2	3	$2/3$	$-1/4$	$1/5$ . . .

Note that we skip any fraction that has already appeared (for example, $1 = 2/2 = 3/3$, etc.). Thus this scheme provides a means of matching each natural number with one and only one rational and, more surprisingly, of matching each rational with one and only one natural number. With Cantor's definition, the conclusion is immediate: There are as many rationals as natural numbers.

At this point, it must have seemed that *every* infinite set was denumerable, that is, capable of being matched with the positive integers. But in 1874, the mathematical world learned otherwise in Cantor's scholarly paper with the prosaic title "Ueber eine Eigenschaft des Inbegriffes aller reellen algebraischen Zahlen" ("On a Property of the Collection of all Algebraic Numbers"). Here Cantor explicitly found a set that was not denumerably infinite.

Scanning the unspectacular-looking title page shown here, one gets little sense of the revolutionary nature of this document. This stands in contrast to radical departures in the arts, which are often conspicuously innovative. Any layman who, in 1874, had visited Paris to see the paintings of Claude Monet would have been struck by the "impressionist" techniques the artist had introduced. Even the casual observer would have seen in Monet's brushwork, in his rendering of light, a significant departure from the canvases of such predecessors as Delacroix or Ingres. Clearly something radical was going on. Yet in this mathematical landmark of the same year, Georg Cantor had set out upon a course every bit as revolutionary. It is just that mathematics on the printed page lacks the *immediate* impact of a radical piece of art.

The non-denumerable set Cantor found was the collection of *all* real numbers. In fact, his 1874 paper showed that no interval of real numbers, regardless of how small its length, could be put into a one-to-one cor-

Ueber eine Eigenschaft des Inbegriffes aller reellen algebraischen Zahlen.

(Von Herrn *Cantor* in Halle a. S.)

Unter einer reellen algebraischen Zahl wird allgemein eine reelle Zahlgrösse ω verstanden, welche einer nicht identischen Gleichung von der Form genügt:

$$(1.) \quad a_0 \omega^n + a_1 \omega^{n-1} + \cdots + a_n = 0,$$

wo $n, a_0, a_1, \cdots a_n$ ganze Zahlen sind; wir können uns hierbei die Zahlen n und a_0 positiv, die Coefficienten $a_0, a_1, \cdots a_n$ ohne gemeinschaftlichen Theiler und die Gleichung (1.) irreductibel denken; mit diesen Festsetzungen wird erreicht, dass nach den bekannten Grundsätzen der Arithmetik und Algebra die Gleichung (1.), welcher eine reelle algebraische Zahl genügt, eine völlig bestimmte ist; umgekehrt gehören bekanntlich zu einer Gleichung von der Form (1.) höchstens soviel reelle algebraische Zahlen ω, welche ihr genügen, als ihr Grad n angiebt. Die reellen algebraischen Zahlen bilden in ihrer Gesammtheit einen Inbegriff von Zahlgrössen, welcher mit (ω) bezeichnet werde; es hat derselbe, wie aus einfachen Betrachtungen hervorgeht, eine solche Beschaffenheit, dass in jeder Nähe irgend einer gedachten Zahl α unendlich viele Zahlen aus (ω) liegen; um so auffallender dürfte daher für den ersten Anblick die Bemerkung sein, dass man den Inbegriff (ω) dem Inbegriffe aller ganzen positiven Zahlen ν, welcher durch das Zeichen (ν) angedeutet werde, eindeutig zuordnen kann, so dass zu jeder algebraischen Zahl ω eine bestimmte ganze positive Zahl ν und umgekehrt zu jeder positiven ganzen Zahl ν eine völlig bestimmte reelle algebraische Zahl ω gehört, dass also, um mit anderen Worten dasselbe zu bezeichnen, der Inbegriff (ω) in der Form einer unendlichen gesetzmässigen Reihe:

$$(2.) \quad \omega_1, \omega_2, \cdots \omega_\nu, \cdots$$

Cantor's 1874 paper containing his first proof of the non-denumerability of the continuum (photograph courtesy of The Ohio State University Libraries)

respondence with **N**. His original proof took him into the realm of analysis and required some relatively advanced mathematical tools. However, in the year 1891, Cantor returned to this problem and provided the very simple proof that we shall examine.

FIGURE 11.1

Great Theorem: The Non-Denumerability of the Continuum

Here a "continuum" means an interval of real numbers, and we introduce the notation (Figure 11.1)

$$(a,b) = \text{the set of all real numbers } x \text{ such that } a < x < b$$

In the proof below, the particular interval whose non-denumerability we shall establish is (0,1), the so-called "unit interval." Note that real numbers in this interval can be expressed as infinite decimals. For instance,

$$\frac{1}{2} = .50000000\ldots,\ \frac{3}{11} = .27272727\ \ldots, \text{ and } \frac{\pi}{4} = .78539816\ldots$$

For technical reasons, we want to be careful to avoid two different decimal representations of the same number. For instance .50000 . . . = ½ can also be written as .49999999 . . . In such cases, we shall opt for the expansion ending in a string of zeros rather than a string of nines so that, under this convention, the decimal representation of any real number in (0,1) is unique.

We can now take a look at Cantor's proof that (0,1) is not denumerable. His was a *reductio ad absurdum* argument, beginning with the assumption that **N** and (0,1) can be matched in a one-to-one fashion and from this deriving a logical contradiction. It certainly ranks as a very great theorem.

THEOREM The interval of all real numbers between 0 and 1 is not denumerable.

PROOF We assume that the interval (0,1) *can* be matched in a one-to-one fashion with **N**, and from this assumption we shall eventually derive a contradiction. In order to clarify Cantor's reasoning, we exhibit such a supposed correspondence as follows:

N		real numbers in (0,1)
1	$\leftrightarrow$	$x_1 = .371652\ldots$
2	$\leftrightarrow$	$x_2 = .500000\ldots$
3	$\leftrightarrow$	$x_3 = .142678\ldots$
4	$\leftrightarrow$	$x_4 = .000819\ldots$
5	$\leftrightarrow$	$x_5 = .987676\ldots$
$\vdots$	$\vdots$	$\vdots$
n	$\leftrightarrow$	$x_n = .a_1a_2a_3a_4a_5\ldots a_n\ldots$
$\vdots$	$\vdots$	$\vdots$

If this really were a one-to-one correspondence, then every real number in (0,1) would appear somewhere in the right-hand column, matched with a particular natural number on the left.

Cantor now described a real number b, whose decimal expansion $.b_1b_2b_3b_4b_5\ldots b_n\ldots$ we determine as follows:

> Choose b_1 (the first decimal place of b) to be any digit different from the first decimal place of x_1 and simultaneously not equal to 0 or 9.
>
> Choose b_2 (the second decimal place of b) to be any digit different from the second decimal place of x_2 but not equal to 0 or 9.
>
> Choose b_3 (the third decimal place of b) to be any digit different from the third decimal place of x_3 but not equal to 0 or 9.
>
> Generally, choose b_n (the nth decimal place of b) to differ from the nth decimal place of x_n but not equal to 0 or 9.

To help understand this process, refer to the specific matching above. The first decimal place of x_1 is "3," so we can choose $b_1 = 4$; the second decimal place of x_2 is "0," so we can choose $b_2 = 1$; the third decimal place of x_3 is "2," so we take $b_3 = 3$; the fourth decimal place of x_4 is "8," so we take $b_4 = 7$; and so on. Thus, our number b looks like:

$$b = .b_1b_2b_3b_4b_5\ldots = .41378\ldots$$

Now we need only observe two simple, but contradictory, facts:

(1) b is a real number since it is an infinite decimal. Because of our prohibition on choosing 0s or 9s, the number b cannot be $.00000\ldots =$

0 nor can it be .99999 . . . = 1. In other words, b must fall *strictly* between 0 and 1. Consequently, b must appear somewhere in the right-hand column of our list above.

but

(2) b cannot appear anywhere among the numbers $x_1, x_2, x_3, \ldots x_n,$ . . . , for certainly $b \neq x_1$ since b and x_1 differ in their first decimal place; and $b \neq x_2$, since b and x_2 differ in their second decimal place; and generally, $b \neq x_n$ since b and x_n differ in the nth decimal place.

Thus, while (1) tells us that b *must* be in the right-hand column, (2) simultaneously tells us it *cannot* be there since it has been "designed" explicitly not to match any of the numbers $x_1, x_2, \ldots, x_n,$ and so on. This logical impasse shows that our original assumption, namely, that a one-to-one correspondence exists between the natural numbers and *all* real numbers in the unit interval, was invalid. By contradiction, we must conclude that such a correspondence is *impossible,* and consequently the set of all real numbers between 0 and 1 is not denumerable.

Q.E.D.

There is an additional reason for avoiding "9s" when selecting the digits of b. Reconsider the specific correspondence introduced above, but this time allow 9s to be used as the values of b_n (provided, of course, they differ from the nth decimal place of x_n). Then we could choose $b_1 = 4$, $b_2 = 9$, $b_3 = 9$, $b_4 = 9$, and so on. As a consequence, our resulting number is $b = .49999 \ldots$. This, however, is precisely ½, a number that *does* appear in the right-hand column as x_2. The contradiction we sought—of determining a real number b not contained on the right-hand side—has thus vanished. But by taking the precaution of avoiding "9" in the construction of b, we eliminate the technical pitfall created by this dual representability of infinite decimals and the proof stands.

Cantor himself was obviously quite pleased with this argument, which he called ". . . remarkable . . . because of its great simplicity." Note that he focused on those decimal places along the array's descending diagonal—the first place of the first real number, the second place of the second real number, and so on. This technique thus acquired the name of Cantor's "diagonalization process."

It is essential to observe that our argument did not depend on the specific matching we introduced above for the purposes of illustration. The very same reasoning would show that no one-to-one correspondence is possible.

Skeptical students often concede that Cantor found a number b not appearing on the original list but suggest the following remedy: Why not simply place *b* opposite the natural number 1, and then move each number on the list down one position? In this way, 2 would match with x_1, 3 would match with x_2, and so on. The contradiction Cantor reached would seem to have been eliminated, since *b* now appears atop the right-hand column.

Unfortunately for the skeptic, Cantor could sit quietly until these adjustments in the list were completed and then, by repeating the diagonal process with the *new* list, create a real number b' that appears nowhere on it. If our skeptic inserted b' at the beginning, we could diagonalize again to get a missing b''. In short, a one-to-one matching between **N** and (0,1) is impossible. The skeptics must become believers.

Thus Cantor had shown that many infinite sets—particularly the rational numbers—had cardinality $\aleph_0$, but that the interval of real numbers between 0 and 1, while infinite, was somehow "more" infinite. Points in this interval are so abundant that they outnumber the positive integers.

In this sense, there is nothing special about the unit interval (0,1). Given *any* finite interval (a,b), we could introduce the function $y = a + (b - a)x$, which is a one-to-one correspondence matching the points of (0,1)—that is, the *x*s—with the points of (a,b)—the *y*s, as shown in Figure 11.2. This one-to-one matching guarantees that the intervals (0,1) and (a,b) have the same (non-denumerable) cardinality. It is perhaps surprising to note that the cardinality of an interval is independent of the interval's length; the set of all real numbers between 0 and 1 has no fewer members than the set of all real numbers between 2 and 1000 (in this case the function $y = 998x + 2$ provides the desired one-to-one correspondence). At first this seems counter-intuitive, but as

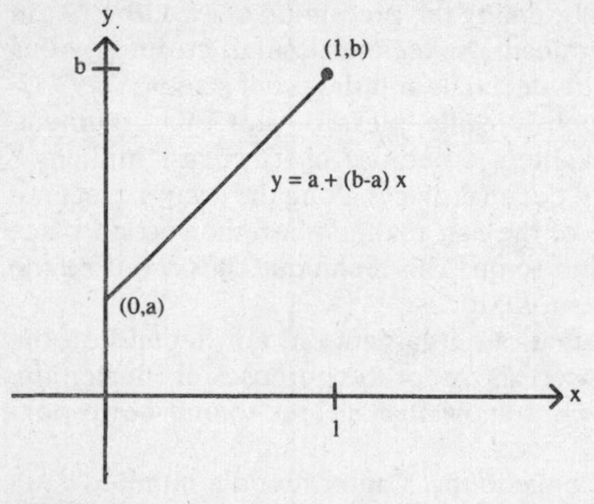

FIGURE 11.2

one gets used to the nature of infinite sets, one soon loses faith in the power of naive intuition.

From here it is a small step to show that the set of *all* real numbers likewise is of the same cardinality as (0,1). This time, a one-to-one matching is given by

$$y = \frac{2x - 1}{x - x^2}$$

As the graph in Figure 11.3 shows, to each x in (0,1) has been associated a unique real number and, conversely, for each real number y, there is one and only one x in (0,1) matched with it. In short, this is the necessary one-to-one correspondence.

We can now follow Cantor's lead and take another bold step. Just as **N** was used as the base set to introduce $\aleph_0$ as the first transfinite cardinal, so the interval (0,1) will be the standard for defining a new, and larger, infinite cardinal. That is, we can define the cardinality of the unit interval to be **c** (a letter standing for "continuum"). Then the preceding discussions show that not only does (0,1) have cardinality **c**, but any interval of finite length, as well as the set of all real numbers itself, has this same cardinality. Further, the non-denumerability of (0,1) shows that **c** is a different cardinal than $\aleph_0$. Cantor was thus on his way to constructing a hierarchy of transfinite numbers.

All of these considerations began to shed light on the intrinsic difference between the set of rationals and the set of irrationals—a difference going far beyond the fact that the former can be expressed as ter-

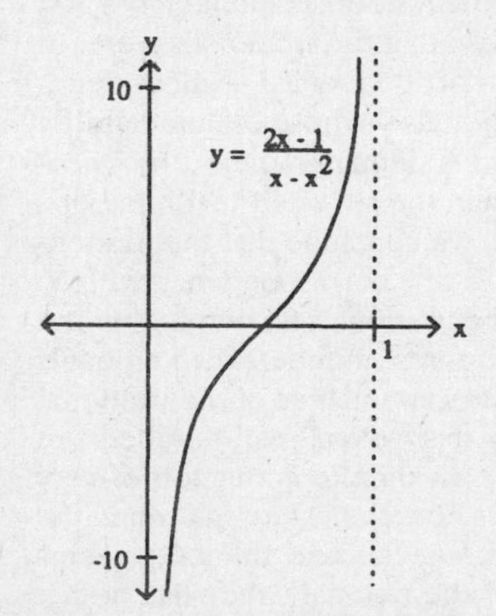

FIGURE 11.3

minating or repeating decimals and the latter can not. To see this more clearly, Cantor needed only one additional result:

THEOREM U If **B** and **C** are denumerable sets and **A** is the set of all elements belonging either to **B** or to **C** (or to both), then **A** itself is denumerable. (In this case, we say **A** is the *union* of **B** and **C** and write **A** = **B** ∪ **C**.)

PROOF The assumed denumerability of **B** and **C** guarantees the individual one-to-one correspondences:

$$\begin{array}{lccccc} \mathbf{N}: & 1 & 2 & 3 & 4 & \ldots \\ & \updownarrow & \updownarrow & \updownarrow & \updownarrow & \\ \mathbf{B}: & b_1 & b_2 & b_3 & b_4 & \ldots \end{array} \quad \text{and} \quad \begin{array}{lccccc} \mathbf{N}: & 1 & 2 & 3 & 4 & \ldots \\ & \updownarrow & \updownarrow & \updownarrow & \updownarrow & \\ \mathbf{C}: & c_1 & c_2 & c_3 & c_4 & \ldots \end{array}$$

By jumping back and forth between the elements of **B** and **C**, we can generate a one-to-one matching between **N** and **A** = **B** ∪ **C**:

$$\begin{array}{lccccccccc} \mathbf{N}: & 1 & 2 & 3 & 4 & 5 & 6 & 7 & 8 & \ldots \\ & \updownarrow & \updownarrow & \updownarrow & \updownarrow & \updownarrow & \updownarrow & \updownarrow & \updownarrow & \\ \mathbf{A}: & b_1 & c_1 & b_2 & c_2 & b_3 & c_3 & b_4 & c_4 & \ldots \end{array}$$

so that **A** itself is denumerable. This shows that the union of two denumerable sets is denumerable.

Q.E.D.

Now we can prove a major difference between the set of rationals and the set of irrationals: we have shown that the former is denumerable and we claim that the latter is not. For, suppose that the irrationals were, in fact, denumerably infinite. Then the union of all rationals—whose denumerability we have proved—and all irrationals—whose denumerability we have just assumed—would likewise be a denumerable set by Theorem U. But this union is nothing other than the set of all real numbers, a non-denumerable set. By contradiction, we conclude that the irrationals are too abundant to be put into a one-to-one correspondence with **N**.

In less formal terms, this means that the irrationals far outnumber the rationals. The reason that there are far more real numbers than rationals can only be explained by the overwhelming abundance of the irrational numbers. Mathematicians sometimes say that "most" real numbers are irrational; the set of rationals, admittedly an infinite collection of very important, densely distributed numbers, is nonetheless just a drop in the bucket. Suddenly a denumerable set among the real numbers seems insignificant, even though, in the case of the rationals, they had at first appeared to be so plentiful. Not so, said Cantor. In terms of cardinality, the rationals are really quite scarce. It is the irrationals that dominate the scene.

These were strange theorems, arising out of the desire to probe the deeper secrets of the calculus. Cantor's work had certainly shed light on the intrinsic differences between sets of real numbers, and this could help explain some hitherto unexplainable phenomena. But if the origins of Cantor's work could be traced back to questions arising from the arithmetization of the calculus, his set theory was about to take on a dramatic life of its own, as we shall see in the next chapter.

Epilogue

All of this was startling enough, but Cantor's 1874 paper contained one more astounding result. Having shown the non-denumerability of intervals, Cantor now applied this fact to a familiar and difficult question that had long exasperated mathematicians—the existence of transcendental numbers.

We have just seen that the set of all real numbers can be subdivided into the relatively scarce rationals and the relatively abundant irrationals. But we recall from the Epilogue to Chapter 1 that real numbers can be split into two different exhaustive and mutually exclusive categories, the algebraic numbers and the transcendental numbers.

The algebraic numbers seem to constitute a vast set. All of the rationals are in it, as are all constructible magnitudes, as well as a multitude of irrationals, such as $\sqrt{2}$ or $\sqrt[3]{5}$. The transcendental numbers, by contrast, are extremely hard to come by. It was Euler who first speculated that transcendental numbers exist—that is, that not all real numbers are of the relatively tame algebraic variety—but the first example of a specific transcendental number was only provided by the Frenchman Joseph Liouville in 1844. When Cantor approached this subject in 1874, Lindemann's proof that π was transcendental still lay almost a decade in the future. In other words, as Cantor was developing his theory of the infinite, there were very few transcendental numbers on the scene. Perhaps they constitute the exception among the real numbers and not the rule.

However, Georg Cantor was becoming accustomed to turning exceptions into rules, and he did so here. He first proved that the set of all algebraic numbers is *denumerable.* Armed with this fact, Cantor was ready to consider the seemingly rare transcendental numbers.

He began with an arbitrary interval (a,b). He had proved that the algebraic numbers contained within this interval formed a denumerable set; if the transcendental numbers likewise were denumerable, then

(a,b) itself would be denumerable by Theorem U. But he had already shown that intervals are not denumerable. This meant that the transcendentals must vastly outnumber the algebraic numbers in any interval!

To put it differently, Cantor knew that there were far more real numbers in (a,b) than could be accounted for by the relatively skimpy collection of algebraic numbers. Where were all these additional real numbers coming from? They must be the transcendentals, which exist in overwhelming abundance.

This was a genuinely provocative theorem. After all, at this point only a few non-algebraic numbers were known. Yet here was Georg Cantor confidently saying that it was the transcendentals that were in the vast majority, and he did so ***without exhibiting a single concrete example*** of a transcendental number! Instead, he "counted" the points of the interval and realized that the algebraic numbers contained within were responsible for only a small part of this count. It was a startlingly indirect way to get at the existence of transcendentals. Eric Temple Bell, a popular writer on the history of mathematics, summarized the situation with the following poetic image:

> The algebraic numbers are spotted over the plane like stars against a black sky; the dense blackness is the firmament of the transcendentals.

Such was the legacy of Cantor's single, landmark 1874 publication. A number of mathematicians, considering these results, shook their heads in amazement or outright skepticism. To mathematical conservatives, the comparisons of infinities seemed like the outrageous, romantic escapade of a young and slightly mystical scholar; asserting the abundance of the transcendentals without producing a single example of one was sheer folly.

Georg Cantor heard these criticisms. But he believed passionately in the course he was following, and he was just getting started. What he had in store would make even these discoveries pale by comparison.

12
Chapter

Cantor and the Transfinite Realm
(1891)

The Nature of Infinite Cardinals

Where did Georg Cantor go from here? In the years following his 1874 paper, Cantor looked ever more closely into the nature of infinite sets of points. His research took many directions and opened unexpected new doors, but it always featured his characteristic boldness and imagination in addressing previously unanswered—indeed, *unasked*—questions about the infinite.

As soon as he realized that he could successfully define more than a single transfinite cardinal, Cantor needed to formalize the concept of "less than" for this new kind of number. For this purpose, it was reasonable to rely again on one-to-one correspondences, but, as should by now be clear, one must proceed with great care.

Before pursuing this matter in the abstract, we should think again about our primitive society in which people could only count to three. Recall that a gifted member of this society had introduced five as the new cardinal number possessed by any set that could be put into a one-to-one correspondence with the fingers of her right hand. Now how might she go about showing that three is less than five? (To us this seems quite trivial, but we are used to counting beyond three). Suppose, after some thought and a good deal of searching, she found a fellow with only three fingers—say, the thumb, index, and ring finger—on his right hand. She then could match *all* of his right-hand fingers with *some* of hers—that is, match their thumbs, index, and ring fingers—in a one-to-one fashion. This would leave two unmatched fingers on her right hand, and this surplus of her fingers shows that five exceeds three.

One could try to extend this definition to general sets by saying that the cardinality of set **A** is less than the cardinality of set **B**—written $\overline{\mathbf{A}} < \overline{\mathbf{B}}$—if there is a one-to-one correspondence from all the elements of **A** to some of the elements of **B**. That is, if **A** can be matched with a *subset* of **B** in a one-to-one manner, then surely **A** would appear to have fewer members than **B**.

Unfortunately, while this definition seems fine for showing that 3 < 5, it is quite unsatisfactory when we move to infinite sets. Consider, for instance, the set of natural numbers, **N**, and the set of rationals, **Q**. It is easy to write down a one-to-one correspondence between all of **N** and a small subset of **Q**, namely, those positive fractions with numerator one:

$$\begin{array}{lcccccccccc}
\mathbf{N}: & 1 & 2 & 3 & 4 & 5 & 6 & \cdots & n & \cdots \\
 & \updownarrow & \updownarrow & \updownarrow & \updownarrow & \updownarrow & \updownarrow & & \updownarrow & \\
\mathbf{Q}: & 1 & \frac{1}{2} & \frac{1}{3} & \frac{1}{4} & \frac{1}{5} & \frac{1}{6} & \cdots & \frac{1}{n} & \cdots
\end{array}$$

Yet we certainly do not want to use this matching to conclude that $\overline{\mathbf{N}} < \overline{\mathbf{Q}}$. In fact, we have already seen that there is a different one-to-one correspondence between all of **N** and *all* of **Q**, so that both sets have the same cardinality. At first glance, we seem to face an unpleasant dilemma.

Cantor deftly found his way out of the quagmire by first introducing not the notion of "less than" but that of "less than or equal to":

□ **Definition:** If **A** and **B** are sets, then we say $\overline{\mathbf{A}} \leq \overline{\mathbf{B}}$ if there exists a one-to one correspondence from all of the points of **A** to a subset of the points of **B**.

Note that the "subset" of the points of **B** may be all the points of **B**, in which case we would have $\overline{\mathbf{A}} = \overline{\mathbf{B}}$. This, of course, is perfectly consistent with the more inclusive statement that $\overline{\mathbf{A}} \leq \overline{\mathbf{B}}$. Furthermore, the matching above between all of **N** and part of **Q** merely shows that $\overline{\mathbf{N}} \leq \overline{\mathbf{Q}}$, which is not contradictory since both sets have cardinality $\aleph_0$.

Now Cantor could define what he meant by a strict inequality between the cardinalities of two sets:

□ **Definition:** $\overline{\mathbf{A}} < \overline{\mathbf{B}}$ if $\overline{\mathbf{A}} \leq \overline{\mathbf{B}}$ (as in the previous definition) but if there is no one-to-one correspondence between **A** and **B**.

On the surface, this may seem utterly trivial, but a little thought shows that it hinges upon important properties of one-to-one correspondences. For, to show that $\overline{\mathbf{A}} < \overline{\mathbf{B}}$ we must first find a one-to-one correspondence between all of **A** and part of **B** (thereby establishing that $\overline{\mathbf{A}} \leq \overline{\mathbf{B}}$), and then show that there can be no one-to-one correspondence between all of **A** and all of **B**. The matter quickly becomes far from self-evident.

Nonetheless, this definition does the job. For instance, it proves that $3 < 5$ for our primitive friends. That is, the matching of thumb, index, and ring finger to the corresponding fingers on the five-fingered right hand shows that $3 \leq 5$; however, there is no way to match all five of her fingers in a one-to-one way with her colleague's three digits, and thus the cardinal numbers 3 and 5 are not equal. The conclusion is then that $3 < 5$.

Moving up to the infinite cardinals, this same logical approach suffices to show that $\aleph_0 < \mathbf{c}$, for, we can easily provide a one-to-one matching between all of **N** and a subset of (0,1):

$$\begin{array}{lccccccccc} \mathbf{N}: & 1 & 2 & 3 & 4 & 5 & \ldots & n & \ldots \\ & \updownarrow & \updownarrow & \updownarrow & \updownarrow & \updownarrow & & \updownarrow & \\ (0,1): & \frac{1}{\pi} & \frac{1}{2\pi} & \frac{1}{3\pi} & \frac{1}{4\pi} & \frac{1}{5\pi} & \ldots & \frac{1}{n\pi} & \ldots \end{array}$$

Hence, $\overline{\mathbf{N}} \leq \overline{(0,1)}$. But Cantor's diagonal proof showed that no one-to-one correspondence exists between these two sets. Thus, $\overline{\mathbf{N}} \neq \overline{(0,1)}$. Together these facts lead us to conclude that $\overline{\mathbf{N}} < \overline{(0,1)}$—that is, $\aleph_0 < \mathbf{c}$.

Cantor now had formulated a method for comparing the sizes of cardinal numbers. Notice that an immediate consequence of this definition is the intuitively pleasing fact that, if **A** is a *subset* of **B**, then $\overline{\mathbf{A}} \leq \overline{\mathbf{B}}$. That is, we can surely match each point of **A** with itself, a one-to-one correspondence between all of **A** and a subset of **B**. Consequently, the cardinality of a set is greater than or equal to the cardinality of any of its subsets. Amid a body of counter-intuitive results, this one seems comforting.

With the ability to compare cardinalities, Cantor introduced an important, and from his viewpoint a very critical, assertion:

$$\text{If} \quad \overline{\overline{\mathbf{A}}} \leq \overline{\overline{\mathbf{B}}} \quad \text{and if} \quad \overline{\overline{\mathbf{B}}} \leq \overline{\overline{\mathbf{A}}} \quad \text{then} \quad \overline{\overline{\mathbf{A}}} = \overline{\overline{\mathbf{B}}}$$

If we restrict our attention to finite cardinals, this result appears quite routine. When dealing with transfinite cardinals, its obviousness vanishes. Think carefully about what Cantor has proposed: If there is a one-to-one matching between all of **A** and part of **B** (that is, $\overline{\overline{\mathbf{A}}} \leq \overline{\overline{\mathbf{B}}}$) and if there is an analogous matching between all of **B** and part of **A** (that is, $\overline{\overline{\mathbf{B}}} \leq \overline{\overline{\mathbf{A}}}$), then we would like to conclude that a one-to-one correspondence exists between all of **A** and all of **B** (that is, $\overline{\overline{\mathbf{A}}} = \overline{\overline{\mathbf{B}}}$). But where does one get this last correspondence? A little consideration reveals that this is a profound assertion indeed.

Georg Cantor was convinced that this statement was true, perhaps indicating his faith in the "reasonableness" of his emerging set theory, yet he never was able to give a satisfactory proof of the result (a sure sign of its complexity). Fortunately, the theorem was proved independently by two mathematicians, Ernst Schröder (in 1896) and Felix Bernstein (in 1898). Because of its joint origins, the result is today known as the "Schröder-Bernstein theorem," although one also finds it called the "Cantor-Bernstein theorem," or the "Cantor-Schröder-Bernstein theorem," or other permutations of these names. Its appellation aside, the theorem is a useful tool in the study of transfinite cardinals.

Although the proof is beyond the scope of this book, we can illustrate the theorem's power by determining the cardinality of the set **I** of all irrational numbers. We have seen in the previous chapter that the irrationals are not denumerable; that is, their cardinality exceeds $\aleph_0$. But we did not precisely determine that cardinality. To do so, we can use the Schröder-Bernstein theorem.

First, note that the irrationals form a subset of the real numbers, so our previous comments guarantee that $\overline{\overline{\mathbf{I}}} \leq \mathbf{c}$. On the other hand, consider the matching that takes each real number to an irrational defined as follows: If $x = M.b_1b_2b_3b_4 \ldots b_n \ldots$ is a real number in its decimal form, where M is its integer part, we associate with x the real number

$$y = M.b_1 0 b_2 11 b_3 000 b_4 1111 b_5 00000 b_6 111111 \ldots$$

That is, we insert one 0 after our first digit, two 1s after the second, three 0s after the third, and so forth. For instance, the real number $x =$ 18.1234567 . . . is matched with

$$y = 18.10211300041111500000611111117 \ldots$$

while the number $x = -7.25 = -7.25000 \ldots$ is matched with

$$y = -7.20511000001111000000111111 \ldots$$

Regardless of which real number x we choose, the resulting y has a decimal expansion that neither terminates nor repeats, since we get ever longer blocks of consecutive 0s and 1s appearing in its expansion. Thus, the matching takes each real number x to an irrational number y.

Moreover, the matching is one-to-one. For if we were given a resulting y, say 5.30411400071111000002 . . . , we could "unravel" it to the one and only possible x from which it came, in this case $x = 5.344712 \ldots$. We should observe that not every irrational number ends up being matched with something. The irrational $y = \sqrt{2} = 1.414159 \ldots$ does not have the correct sequence of 0s and 1s in its decimal expansion to be the mate of any real number x under this matching.

This one-to-one correspondence between *all* of the real numbers and *some* of the irrationals implies that $\mathbf{c} \leq \bar{\mathbf{I}}$. But we have already noted that $\bar{\mathbf{I}} \leq \mathbf{c}$ and so by the Schröder-Bernstein theorem we conclude that the cardinality of the set of irrationals is $\mathbf{c}$, the same as the cardinality of the set of all real numbers.

With this result, posed by Cantor and proved by Schröder and Bernstein, one of the great issues of transfinite cardinals was successfully resolved, but Georg Cantor was never one to run out of questions about his marvelous creation. Another was whether there existed any cardinals greater than $\mathbf{c}$. Judging from his early correspondence, he sensed the answer was "yes" and thought he knew where to look for a more abundant set of points.

To Cantor, the key to finding a larger cardinality than that possessed by the one-dimensional interval (0,1) was to look at the two-dimensional square bounded by (0,1) on the x-axis and (0,1) on the y-axis, as illustrated in Figure 12.1. Writing to his friend Richard Dedekind in January of 1874, Cantor asked whether these two sets, the interval and the square, could be put into a one-to-one correspondence. He was nearly certain that no such correspondence was possible between the two-dimensional square and the one-dimensional segment, for it seemed

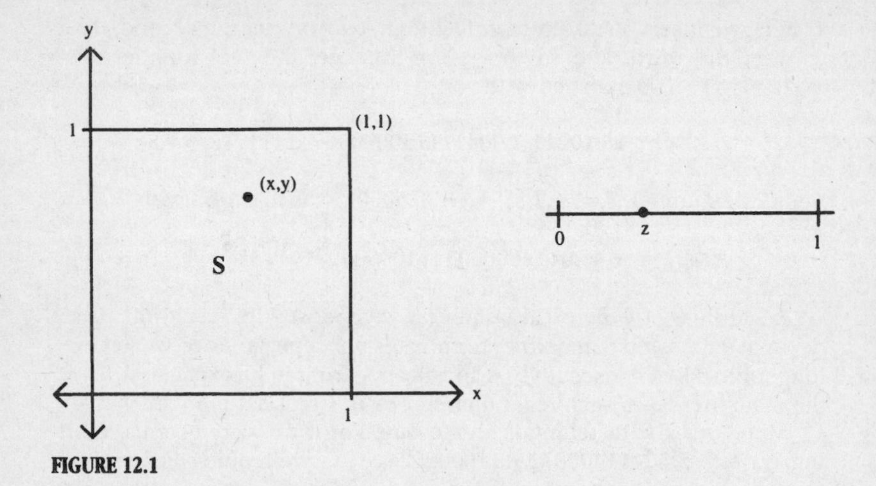

FIGURE 12.1

clear that the former was far more abundant in points. Although constructing a proof might be difficult, Cantor felt that an argument might be "almost superfluous."

Interestingly, this almost superfluous proof never materialized. Try as he might, Cantor failed to establish that it was impossible to match the interval and the square in a one-to-one manner. Then, in 1877, he found that his original intuition was entirely wrong. Such a correspondence does exist!

For a proof of this surprising fact, we shall let **S** denote the square consisting of all ordered pairs (x,y), where $0 < x < 1$ and $0 < y < 1$. It is easy to produce a one-to-one correspondence between *all* of the unit interval and *part* of **S** by simply matching z in (0,1) with the ordered pair $(z,\frac{1}{2})$ in **S**. By our previous definition, we conclude that $\overline{\overline{(0,1)}} \leq \overline{\overline{\mathbf{S}}}$.

On the other hand, for any point (x,y) in **S**, the individual coordinates x and y are themselves infinite decimals. That is, $x = .a_1a_2a_3a_4 \ldots a_n \ldots$ and $y = .b_1b_2b_3b_4 \ldots b_n \ldots$. As we did in Chapter 11, we insist that these decimal expansions be unique—when confronting a number that can be represented as ending in either a string of 0s or a string of 9s, we shall use the former representation rather than the latter, so that we would express $\frac{1}{5}$ by the decimal .2000 . . . rather than the equivalent. 1999 . . .

Having adopted this convention, we associate to each (x,y) in **S** the point z in (0,1) defined by

$$z = .a_1b_1a_2b_2a_3b_3a_4b_4 \ldots$$

For instance, the pair $(2/11, \sqrt{2}/2) = (.18181818\ldots, .70710678\ldots)$ in the square would be matched with .1780178110861788 . . . on the interval simply by shuffling the decimal places together. Nothing could be simpler. Likewise, given any point in the interval that is matched with some point in **S**, we can easily unshuffle to return to the unique ordered pair from which it came. That is, $z = .93440125\ldots$ must have as its predecessor in the square the pair

$$(x,y) = (.9402\ldots, .3415\ldots)$$

We note that under this correspondence not every point in the unit interval is the mate of some point in the square. For instance, the point $z = 6/55 = .109090909\ldots$ in (0,1) would unshuffle to the ordered pair (.19999 . . . , .0000 . . .). But we had ruled out using .1999 . . . at all, choosing instead the equivalent .2000 . . . ; even worse, the second coordinate $.0000\ldots = 0$ does not fall *strictly* between 0 and 1, so the unraveling process has taken us outside of **S**. In other words, 6/55 = .10909090 . . . is not matched with any point in the square.

Nonetheless we have here a one-to-one matching between *all* the points of **S** and *some* of the points of (0,1), and we conclude that $\overline{\overline{\mathbf{S}}} \le \overline{\overline{(0,1)}}$. This fact, coupled with the previous inequality that $\overline{\overline{(0,1)}} \le \overline{\overline{\mathbf{S}}}$, allows us to apply the Schröder-Bernstein theorem to deduce that, indeed, $\overline{\overline{\mathbf{S}}} = \overline{\overline{(0,1)}} = c$.

This discussion shows that, in spite of the difference in dimension, the points of the square are no more abundant than the points in the interval. Both sets have cardinality **c**. To say the least, this was a surprise. Writing to Dedekind in 1877 to report this discovery, Cantor exclaimed, "I see it but I do not believe it!"

So where does one look to find a transfinite cardinal larger than **c**? Cantor could easily show that a larger square, or even all the points in the entire plane, have the same cardinality as the unit interval (0,1). Even going to a three-dimensional cube did not increase the cardinality. It looked as though **c** might be the ultimate transfinite cardinal.

But things proved very much otherwise. In 1891, Cantor succeeded in showing that larger transfinite cardinals exist, and exist in unbelievable profusion. His result is today usually called Cantor's theorem. Given that he proved so many critical theorems in his career, the name given this one indicates the high regard in which it has come to be held. It is a result as stunning as any that set theory is likely to see.

Great Theorem: Cantor's Theorem

To discuss the proof, we need to introduce an additional concept:

□ **Definition:** Given a set **A**, the *power set* of **A**, denoted by $P[\mathbf{A}]$, is the set of all the subsets of **A**.

This seems simple enough. For instance, if $\mathbf{A} = \{a,b,c\}$, then **A** has eight subsets, and the power set of **A** is the set containing these eight subsets, namely:

$$P[\mathbf{A}] = \{\{\ \}, \{a\}, \{b\}, \{c\}, \{a,b\}, \{a,c\}, \{b,c\}, \{a,b,c\}\}$$

Note that the empty set, { }, and the set **A** itself are two of the elements of the power set; this is true regardless of which set **A** we consider. Note also that the power set is itself a *set.* This elementary fact is sometimes easy to overlook but played a key role in Cantor's thinking.

Clearly, for our example above, the power set has cardinality greater than that of the set itself. That is, where **A** contains 3 elements, its power set contains $2^3 = 8$ members. It is not hard to show that a 4-element set has $2^4 = 16$ subsets; a 5-element set has $2^5 = 32$ subsets; and generally an n-element set **A** has 2^n subsets. We can express this symbolically by $\overline{P[\mathbf{A}]} = 2^n$.

But what happens if **A** is an infinite collection? Are infinite sets likewise outnumbered by their power sets? It was Cantor's theorem that answered this provocative question:

THEOREM If **A** is *any* set, then $\overline{\mathbf{A}} < \overline{P[\mathbf{A}]}$.

PROOF To establish this result, we must rely on Cantor's definition of strict inequality between transfinite cardinals, as introduced earlier in this chapter. It is clear that we can easily find a one-to-one correspondence between **A** and a *part of* $P[\mathbf{A}]$ For, if $\mathbf{A} = \{a, b, c, d, e, \ldots\}$, we can match the element a with the subset $\{a\}$, the element b with the subset $\{b\}$, and so on. Of course, these subsets $\{a\}, \{b\}, \{c\}, \ldots$ constitute just a small portion of the collection of *all* subsets of **A**, and so this one-to-one matching guarantees that $\overline{\mathbf{A}} \leq \overline{P[\mathbf{A}]}$.

That much was easy. It remains to show that **A** and $P[\mathbf{A}]$ do not have the same cardinality. To begin an indirect proof of this, we suppose the opposite and derive a contradiction. That is, assume there exists a one-to-one correspondence between all of **A** and *all of* $P[\mathbf{A}]$. In order to follow the argument from here, we would do well to introduce an example of such a supposed correspondence for the sake of later reference:

elements of **A**		elements of $P[\mathbf{A}]$ (i.e., subsets of **A**)
a	$\longleftrightarrow$	$\{b,c\}$
b	$\longleftrightarrow$	$\{d\}$
c	$\longleftrightarrow$	$\{a,b,c,d\}$
d	$\longleftrightarrow$	$\{\ \}$
e	$\longleftrightarrow$	**A**
f	$\longleftrightarrow$	$\{a,c,f,g,\ldots\}$
g	$\longleftrightarrow$	$\{h,i,j,\ldots\}$
.		.
.		.
.		.

This, then, is a hypothesized one-to-one correspondence matching all the elements of **A** with all of the elements of $P[\mathbf{A}]$. Note that, under this matching, some elements of **A** belong to the subset with which they are matched; for instance, c is a member of the set $\{a, b, c, d\}$ with which it is matched. On the other hand, some elements of **A** do not belong to their matching subsets; for instance, a is not a member of its mate $\{b,c\}$.

Strangely, this dichotomy provides the key to reaching the proof's contradiction, for we now define the set **B** as follows:

> **B** is the set of each and every element of the original set **A** that *is not* a member of the subset with which it is matched.

Referring to the illustrative correspondence above, we see that a belongs to **B**, as do b (since it is not an element of $\{d\}$), d (which certainly is not in the empty set), and g (not a member of $\{h,i,j,\ldots\}$). However c, e, and f fail to qualify as members of **B** since each is a member of, respectively, $\{a,b,c,d\}$, **A** itself, and $\{a,c,f,g\ldots\}$.

In this manner, the set $\mathbf{B} = \{a,b,d,g,\ldots\}$ is generated. Of course, at its most basic **B** is simply a *subset* of the original set **A**. Consequently, **B** belongs to the power set of **A** and thus must appear somewhere in the right-hand column of the matching shown. But we began by assuming that we had a one-to-one correspondence, and so we conclude that, in the left-hand column, there must be some element y in **A** that is matched with **B**:

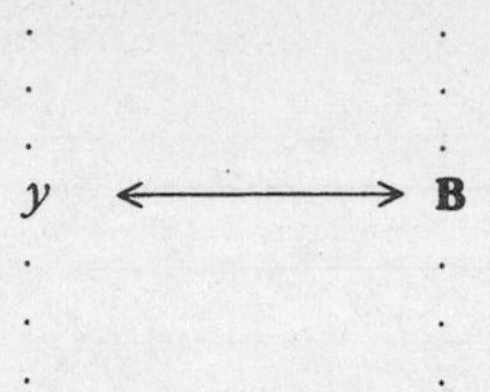

So far, so good. But now we ask the fatal question: "Is y an element of **B**?" There are, of course, two possibilities:

CASE 1 Suppose y *is not* an element of **B**.

Then, by our initial definition of **B** as ". . . each and every element of the original set **A** that is *not* a member of the subset with which it is matched," we see that y must indeed be granted membership in **B**, for y is, in this case, not a member of the set to which it is matched.

In other words, if we begin by supposing that y is not in **B**, we are forced to conclude that y must be made a member of **B**. This is a clear contradiction, and we reject Case 1 as impossible.

Case 2 Suppose y *is* an element of **B**.

Again we refer to the membership criterion for **B**. Since Case 2 assumes that y is in **B**, then y must meet the membership criterion; that is, y is not an element of the set to which it is matched. Alas, the set to which y is matched is **B**, and so y cannot be a member of set **B**.

So, beginning with the assumption of Case 2 that y is in **B**, we are immediately forced to conclude that it is not. Again, a logical impasse has appeared.

Something is terribly wrong. Cases 1 and 2, the only cases possible, both lead to contradictions. We conclude that somewhere in the argument there lies an erroneous assumption. The problem, of course, is that we assumed at the outset that there was a one-to-one correspondence between **A** and $P[\mathbf{A}]$. Our contradiction has clearly destroyed this assumption: no such correspondence can exist.

Finally, combining our conclusions that $\overline{\overline{\mathbf{A}}} \leq \overline{\overline{P[\mathbf{A}]}}$ but $\overline{\overline{\mathbf{A}}} \neq \overline{\overline{P[\mathbf{A}]}}$, we have have proved Cantor's Theorem: for any set **A**, $\overline{\overline{\mathbf{A}}} < \overline{\overline{P[\mathbf{A}]}}$.

Q.E.D.

Perhaps a concrete example using a finite set will show Cantor's genius in action. Let $\mathbf{A} = \{a,b,c,d,e\}$ and set up a matching between points of **A** and some members of its power set:

elements of **A**		elements of its power set (i.e., subsets of **A**)
a	⟷	$\{a,c\}$
b	⟷	**A**
c	⟷	$\{a,e\}$
d	⟷	$\{d\}$
e	⟷	$\{a,b,c,d\}$

Recalling our definition of **B** as those points in **A** that do not belong to the set with which they are matched, we see that **B** = $\{c,e\}$.

Cantor's critical observation is that **B** *cannot appear in the right-hand column above*, for the logic shows that there is no element to which it could possibly be matched. The wonder of Cantor's proof is that, for any proposed matching between **A** and P[**A**], he cleverly described a member of the power set—namely, **B**—that cannot possibly be matched with any element of **A**. This instantly refutes the possibility of a one-to-one correspondence between a set and its power set.

We need to pause and consider the implications of Cantor's theorem. He proved that, no matter what set one takes initially, its power set has strictly greater cardinality. In his own words:

> . . . in place of any given set **L** another set **M** can be placed which is of greater power [cardinality] than **L**.

Thus, to find the long-sought example of a set with cardinality greater than **c**, we do not look at squares in the plane or cubes in three-dimensional space. Instead, we take the set $P[(0,1)]$, the set of all subsets of points in the interval (0,1). By Cantor's theorem, $\mathbf{c} = \overline{\overline{(0,1)}} < \overline{\overline{P[(0,1)]}}$, and we have found a larger transfinite cardinal.

But now recall the essential fact that a power set is, at its most fundamental, just a set. Thus, the process can be repeated by considering the power set of $P[(0,1)]$, that is, the set of all subsets of the set of all subsets of (0,1). While this is surely a mind-boggling collection, the proof above shows that $\overline{\overline{P[(0,1)]}} < \overline{\overline{P[P[(0,1)]]}}$.

With this genie out of the bottle, there was no stopping Georg Cantor. For we clearly can repeat the process indefinitely, thereby generating an increasing chain:

$$\aleph_0 < \mathbf{c} < \overline{\overline{P[(0,1)]}} < \overline{\overline{P[P[(0,1)]]}} < \overline{\overline{P[P[P[(0,1)]]]}} < \ldots$$

There is barely time to catch one's breath. Not only did Georg Cantor open the door to a first transfinite cardinal ($\aleph_0$) and then to an even larger size of infinity (**c**), but this theorem, applied repeatedly, guaranteed a never-ending chain of ever larger transfinite numbers. There was no end to them.

It is an understatement to say that this conclusion, along with all of Cantor's profound results about the infinite, generated an outcry of opposition. Surely, he had pushed mathematics into unexplored territory where it began to merge into the realms of philosophy and metaphysics. It is worth noting that the metaphysical implications of his mathematics were not lost upon Georg Cantor. According to Joseph Dauben, his foremost modern biographer, Cantor came to find a religious significance in his theory of transfinites, and regarded himself "not only as God's messenger, accurately recording, reporting, and transmitting the newly revealed theory of the transfinite numbers but as God's ambassador as well." Cantor himself wrote:

> I entertain no doubts as to the truths of the transfinites, which I recognized with God's help and which, in their diversity, I have studied for more than twenty years; every year, and almost every day brings me further in this science.

As this passage suggests, religion became a focus of much of Cantor's thought. We recall the mixed religious backgrounds of his parents and can only imagine the rich diversity of theological discussions that must have filled the Cantor household. Perhaps this heightened his interest in such matters. In any case, religious concerns would color much of his thinking, be it in mathematics or in other pursuits.

Such an attitude on the part of this strange, mystical man did little to endear him to his critics. Those who objected to his radical theory of the infinite could advance an *ad hominem* argument against an individual who proclaimed his mathematics to be a message from God. Cantor probably did not help his image when, to this fascination with theological questions, he added a fervent interest in proving that Francis Bacon wrote the works of Shakespeare. This may have struck colleagues as odd, but when he claimed to have uncovered information about the first British king that "will not fail to terrify the English government as soon as the matter is published," a number of eyebrows must have been raised. It was getting hard not to regard Georg Cantor as some sort of kook.

Yet there remained his mathematics. Conservative elements in his native Germany and elsewhere vociferously objected to his work, and bad feelings developed between Cantor and some very influential mathematicians. Certainly these objections were not all of the reactionary, knee-jerk variety, for Cantor's mathematics raised genuinely baffling questions that deeply troubled even mathematicians of good will. One such question is considered in the Epilogue.

Among Cantor's critics was Leopold Kronecker (1823–1891), a powerful figure in the German mathematical community and a fixture at the acclaimed University of Berlin, the institution that had nurtured the famous Weierstrass and his illustrious students (including Cantor himself). Cantor had spent his professional career at the University of Halle, a far less prestigious institution than Berlin, where he longed for an appointment. He keenly felt the slight of being relegated to a lesser university and often attributed the state of affairs to Kronecker's persecution. Attacks flew back and forth between Cantor and his opponents, with the former exhibiting fairly clear paranoid tendencies. In the process, Cantor managed to offend friend and foe alike, which hardly improved his chances of employment at Berlin.

It may come as no surprise that Georg Cantor, living such a life of disappointment and grappling with the most arcane concepts of the infinite, suffered a number of bouts with mental illness. His first breakdown came in 1884, when he was feverishly at work on a result known as the "continuum hypothesis," to be examined shortly. A popular view holds that the stress of his mathematics, coupled with the persecution of Kronecker and others, were responsible for his collapse. Modern analysis of the medical data rejects this as being overblown, for there are suggestions that Cantor exhibited a bipolar (that is, manic-depressive) psychosis, and breakdowns most likely would have occurred in any case. His attacks of mental illness may have been triggered by personal and mathematical difficulties, but they appear to have been of a deeper, more fundamental nature.

Be that as it may, the bouts of instability continued and became more frequent. After a brief hospitalization in 1884, Cantor recovered but remained deeply concerned that the disease could return. Amid his disappointments, mathematical and professional, there came a terrible blow with the unexpected death of his beloved son Rudolf in 1899. Cantor was back in the neuropathic hospital in Halle in 1902, and again in 1904, 1907, and 1911. Often his institutionalizations were followed, upon discharge, by periods of sitting at home immobile and silent.

Cantor's was certainly a troubled life. His death, on January 6, 1918, came while he was again hospitalized for his mental affliction. It was a sad end for a great mathematician.

Looking back on the life and works of Georg Cantor, it is tempting to compare him to his contemporary from the world of art, Vincent Van Gogh. The two men had a certain physical resemblance. Cantor's father was highly religious, and Van Gogh's was a Dutch clergyman. Both were much drawn to artistic enterprises, enjoyed literature, and wrote poetry. We recall that Van Gogh, like Cantor, had an erratic, volatile personality that eventually alienated even friends such as Paul Gauguin. Both men were extremely intense, exhibiting a tremendous devotion to their cho-

Georg Cantor (photograph courtesy of The Ohio State University Libraries)

sen work. And, of course, both men suffered from mental problems that not only saw them institutionalized but also weighed upon their minds as they contemplated recurring attacks in the future.

Most of all, both Van Gogh and Cantor were revolutionaries. Just as Vincent in his brief and turbulent career managed to carry art beyond its impressionist boundaries, so too did Cantor move mathematics in profoundly new directions. Whatever is said about this great and troubled man, we cannot help but admire his courage in exploring the nature of the infinite in an absolutely original way.

Cantor himself, despite his problems, never despaired over the value of his work. Discussing his controversial view of the infinite, he wrote:

> This view, which I consider to be the sole correct one, is held by only a few. While possibly I am the very first in history to take this position so explicitly, with all of its logical consequences, I know for sure that I shall not be the last!

Indeed, he was not. Generations of mathematicians had probed the age-old questions of geometry, algebra, and number theory, but Georg Cantor opened up new and unexpected vistas. Because he both asked and answered questions never before contemplated, it is perhaps fitting that his work has been called the first *truly* original mathematics since the Greeks.

Epilogue

We have alluded to certain matters of set theory that even Cantor's great genius left unresolved. One of the most perplexing of these was the appearance of inexplicable paradoxes—logicians use the term "antinomies"—that arose from Cantor's discoveries. Perhaps the simplest of these logical quagmires follows instantly from Cantor's theorem.

Suppose we collect together the set of all sets, and call it **U** (for "universal set"). This is an inconceivably vast assemblage. It contains all sets of ideas, all sets of numbers, all sets of subsets of numbers, etc. Within **U** we would find each and every set that exists. In this sense, **U** cannot possibly be enlarged; it already contains all possible sets.

But now we apply Cantor's Theorem to **U**. Cantor had proved that $\overline{\overline{\mathbf{U}}} < \overline{\overline{P[\mathbf{U}]}}$, which obviously implies that $P[\mathbf{U}]$ is overwhelmingly more vast than **U** itself. Here we have a contradiction appearing at the very core of Cantorian set theory.

Cantor became aware of such antinomies in 1895, and over the next decades the mathematical community tried to find a way to patch up the logical breach they had created. The final resolution of this affair required the formal axiomatization of set theory—even as Euclid had provided his axiomatic approach to geometry—in which the carefully chosen axioms prohibited just such paradoxes as these. Logically, this was no easy matter. But, in the end, the newly created "axiomatic set theory" more carefully controlled precisely what was and what was not a "set." Under this system, the "universal set" was not a set at all; it was excluded from the collection of objects that the axioms of set theory addressed. Thus, almost by magic, the paradox dissolved.

This resolution was, obviously, a compromise measure, an axiomatic attempt to carve away, with surgical precision, the troubling features of set theory while retaining all of the good points of Cantor's creation. Cantor's own, more informal approach is now called "naive set theory," to contrast it with the logical superstructure of axiomatic set theory. The latter now stands as a satisfactory, albeit rather abstruse and technical, foundation for the theory of sets. It represents a triumph of the sentiments expressed by mathematician David Hilbert, who vowed, "No one will expel us from the paradise that Cantor has created."

But there was another problem that Cantor had failed to resolve satisfactorily, and this concerned him at least as much as the appearance of paradoxes. In fact, it occupied Cantor's attention year after year and is cited by some as playing a significant role in his periods of mental collapse. The result is now known as Cantor's "continuum hypothesis."

It is quite simple to state. The continuum hypothesis asserts that there is no transfinite cardinal falling *strictly* between $\aleph_0$ and **c**. In this sense, it suggests that the cardinals $\aleph_0$ and **c** behave much like the whole numbers 0 and 1. These are the first two finite integers, and no other whole numbers can be inserted between them. Cantor's hypothesis surmised that an analogous role was played by his two transfinites.

Put another way, the continuum hypothesis stated that any infinite subset of real numbers is either denumerable (in which case it has cardinality $\aleph_0$) or can be put into a one-to-one correspondence with (0,1) (in which case it has cardinality **c**). There is no intermediate possibility.

Cantor wrestled with this problem incessantly throughout much of his mathematical career. One of his great assaults upon it came in 1884, the year of his first nervous collapse. In August of that year, Cantor felt that his efforts had succeeded, and he wrote to a colleague, Gustav Mittag-Leffler, that he had proved it. But three months later, he wrote a follow-up letter not only retracting his proof of August but also claiming he now had a proof that the continuum hypothesis was false. This radical shift in his views lasted one short day, after which he again wrote Mittag-Leffler conceding that both of his proofs had been flawed. Acknowledging mathematical errors not once but twice, Cantor still had no idea whether his continuum hypothesis was true.

If Cantor had proved this hypothesis, he could, for instance, have easily determined the cardinality of the transcendental numbers, mentioned in the Epilogue of the previous chapter. As shown there, the transcendentals formed a *non-denumerable* subset of the reals and thus would be forced to have cardinality **c**. It would all have been so easy if only Cantor could have proved his continuum hypothesis.

But he never did. In spite of Herculean efforts on his part, he went to his grave no closer to proving the result than he had been decades earlier. It became perhaps his life's greatest obsession and greatest frustration.

It was not just Georg Cantor who sought the answer. In 1900, Hilbert looked across the broad spectrum of unanswered mathematical problems and identified 23 of them as the critical challenges for mathematicians in the century ahead. First on the list was Cantor's continuum hypothesis, which Hilbert called a ". . . very plausible theorem, which, nevertheless, in spite of the most strenuous efforts, no one has succeeded in proving."

Much more strenuous effort would be expended before some light began to be shed on this simple-looking theorem of set theory. The great breakthrough came in 1940 from the pen of one of the twentieth century's most extraordinary mathematicians, Kurt Gödel (1906–1978). Gödel, using the axiomatized version of set theory, proved that the continuum hypothesis is logically consistent with the other axioms of the theory. That is, there was no way, beginning with the set-theoretic axioms, to *disprove* the continuum hypothesis. Had Cantor been alive, this discovery would have cheered him immeasurably, for it seemed to indicate that he was on the right track.

Or did it? Gödel's result certainly did not prove the hypothesis. The question remained an open one until 1963. Then the mathematician Paul Cohen (1934–) of Stanford University showed that, beginning with the axioms of set theory, we could not *prove* the continuum hypothesis either. Combined with Gödel's work, this settled the question of the continuum hypothesis in a most surprising way: it was simply independent of the other principles of set theory.

This should strike a distant, but familiar, bell. Over two thousand years before, Euclid had introduced his parallel postulate, and subsequent generations expended untold effort in trying to derive it from geometry's other postulates. We subsequently learned that this was an impossible quest, for the parallel postulate is independent of these other principles; it can neither be established nor refuted, but stands apart, like an offshore island.

Cantor's continuum hypothesis occupies an analogous position in the world of set theory. Its adoption becomes a matter of choice, not of necessity, based on the tastes of the mathematician in question. If we wish to explore a set theory where no transfinite cardinals fall between $\aleph_0$ and $\mathbf{c}$, we are perfectly welcome to take the continuum hypothesis as a postulate and thereby fulfill our wish. If instead we prefer a different approach, we are likewise welcome to reject the continuum hypothesis. The parallel with Euclidean and non-Euclidean geometry is striking. This situation provides a remarkable link between one of our century's most famous problems and a classic from ancient Greece. It suggests that, even in mathematics, the more things change, the more they remain the same.

And what of Georg Cantor's unresolved quest to prove the continuum hypothesis? In the light of Gödel's and Cohen's work in the twentieth century, we see that he faced not a difficult task but a hopeless one. This fact stands as a poignant, ironic postscript to the life of this troubled mathematician.

Still, his failure in no way diminishes the legacy of Georg Cantor. We leave him with his own assessment, from 1888, of his bold journey into the transfinite realm:

> My theory stands as firm as a rock; every arrow directed against it will return quickly to its archer. How do I know this? Because I have studied it from all sides for many years; because I have examined all objections which have ever been made against the infinite numbers; and above all because I have followed its roots, so to speak, to the first infallible cause of all created things.

Afterword

With Cantor's transfinite cardinals roaring off to an infinitude of infinities, we finish our tour of great mathematical masterpieces. It has been a long journey—from Hippocrates of Chios to the twentieth century—but, I hope, an impressive one, with a remarkable cast of characters crossing our stage and performing brilliantly in the process. It is a story well worth the telling.

G. H. Hardy, who appeared in the discussion of Ramanujan in Chapter 4, had a keen sense of the aesthetics of mathematical proof. Hardy contended that truly great theorems possess the three characteristics of *economy*, *inevitability*, and *unexpectedness*. I think these properties are well represented among the results we have examined. Euclid's proof of the infinitude of primes was as concise, elegant, and "economical" as anyone could ask. Johann Bernoulli's array of infinite series led inevitably to the divergence of the harmonic series, so that, as was said of Archimedes' mathematics, "once seen, you immediately believe you would have discovered it." And many of our propositions were extremely unexpected, from the fact that lunes are quadrable, to the fact that cubics are solvable, to just about anything done by Georg Cantor. All in all, I hope that Hardy would have approved my selection of "great theorems."

For a valedictory, I offer two quotations, separated by fifteen centuries yet somehow conveying much the same idea. The first comes from the Greek commentator Proclus of the fifth century:

> This, therefore, is mathematics: she gives life to her own discoveries; she awakens the mind and purifies the intellect; she brings light to our intrinsic ideas; she abolishes the oblivion and ignorance which are ours by birth.

Finally, I offer another observation from the twentieth century's Bertrand Russell, whose words began this book's Preface. Russell recognized beauty in mathematics and characterized it about as well as anyone could. I conclude with his comment, which I hope describes the reader's reaction to these mathematical masterpieces.

> Mathematics, rightly viewed, possesses not only truth, but supreme beauty— a beauty cold and austere, like that of sculpture, without appeal to any part of our weaker nature, without the gorgeous trappings of painting or music, yet sublimely pure, and capable of a stern perfection such as only the greatest art can show.

Chapter Notes

Preface

Page v: excerpt from Bertrand Russell, Russell, p. 50. Page vii: excerpt from Hermann Hankel, Kline, p. 200. Page vii: Oliver Heaviside: Logic can be patient . . ., Kline, p. 3.

Chapter 1: Hippocrates' Quadrature of the Lune

Page 2: There is a tradition . . . , Kline, p. 20. Page 3: excerpt from "Moscow Papyrus," Eves, p. 41. Page 5: the existence of a clay tablet, Eves, pp. 27–30. Page 6: Plutarch: . . . at that time Thales, Plutarch, p. 99. Page 6: excerpt from Plutarch, Plutarch, p. 100. Page 11: Rather, we have a summary by Simplicius, Heath, 1981, Vol. 1, pp. 183–184. Page 18: He may well have *thought* he could prove it, Kline, p. 41. Pages 18–19: Hippocrates' quadrature of the lune, Heath, 1981, Vol. 1, p. 185. Page 20: Proclus: . . . squared the lune, Proclus, p. 167. Page 20: Nonetheless, Simplicius, Heath, 1981, Vol. 1, p. 184.

Chapter 2: Euclid's Proof of the Pythagorean Theorem

Page 28: Heath: He was a *man of science*, Heath, 1981, Vol. 1, p. 323. Page 29: Archimedes: any cone is one third part, Heath, 1953, p. 2. Page 30: excerpt from

Carl Sandburg, Sandburg, p. 473. Page 31: excerpt from Bertrand Russell, Russell, p. 37. Page 36: excerpt from Russell, Russell, p. 38. Page 38: excerpt from Russell, Kline, p. 1005. Page 39: Recognizing the danger here, Hilbert, p. 15. Page 42, We are told that the Epicureans, Heath, 1956, Vol. 1, p. 287. Pages 48–50: Euclid's proof of the Pythagorean theorem, Heath, 1956, Vol. 1, pp. 349–350. Pages 51–52: Euclid's proof of the converse, Heath, 1956, Vol. 1, pp. 368–369. Page 53: excerpt from Richard Trudeau, Trudeau, p. 97. Page 53: Proclus, This ought even, Heath, 1956, Vol. 1, p. 202. Page 55: first two excerpts from Gauss, Wolfe, p. 46; third excerpt from Gauss, Fauvel and Gray, p. 498. Page 56: excerpt from Wolfgang Bolyai, Fauvel and Gray, p. 527. Page 56: Johann Bolyai: Out of nothing, Wolfe, p. 44. Page 56: excerpt from Gauss, Wolfe, p. 52. Page 57: excerpt from Wolfgang Bolyai, Wolfe, p. 45. Page 57: excerpt from Georg Friedrich Bernhard Riemann, Wolfe, p. 7.

Chapter 3: Euclid and the Infinitude of Primes

Page 73: Hardy: . . . as fresh and significant, Hardy, 1967, p. 92. Pages 73–74: Euclid's proof of the infinitude of primes, Heath, 1956, Vol. 2, p. 412. Page 77: excerpt from Archimedes, Heath, 1953, p. 2. Page 78: By Euclid's day, five such solids, Heath, 1956, Vol. 3, pp. 438–439. Page 79: Plato: We must think of the individual units, Fauvel and Gray, p. 78. Page 81: Sir Thomas Heath: . . . is and will doubtless remain, Heath, 1981, Vol. 1, p. 358. Page 83: Schnirelmann proved in 1931, Eves, p. 437.

Chapter 4: Archimedes' Determination of Circular Area

Page 85: Archimedes is said to have invented, Eves, p. 126. Pages 85–88: excerpts from Plutarch, Plutarch, pp. 376–380. Pages 92–95: Archimedes' determination of circular area, Heath, 1953, pp. 91–93. Page 95: Archimedes: Since then the area, Heath, 1953, p. 93. Page 95: excerpt from Plutarch, Plutarch, p. 378. Page 99: the second proposition is out of place, Heath, 1981, Vol. 2, p. 50. Page 102: excerpt from Archimedes, Heath, 1953, p. 1. Page 105: excerpt from Plutarch, Plutarch, p. 378. Page 105: Cicero: a small column that emerged, Cicero, Vol. 2, p. 138. Pages 106–109, 111–112: Milestones from the history of π: Eves, pp. 85–90. Page 110: G. H. Hardy: . . . must be true, because if they were not, Hardy, 1959, p. 9.

Chapter 5: Heron's Formula for Triangular Area

Page 115: In the Egyptian town of Syene, Heath, 1981, Vol. 2, pp. 106–107. Page 116: Sir Thomas Heath: a surprisingly close approximation, Heath, 1981, Vol. 2, p. 107. Page 117: We shall go with Howard Eves, Eves, p. 133. Page 118: Heron: Why does a stick break sooner, Heath, 1981, Vol. 2, p. 352. Page 118: Mathematicians had long known, Heath, 1981, Vol. 2, p. 317. Pages 121–126: Heron's proof of his formula, Heath, 1981, Vol. 2, pp. 321–323. Page 127: In an old Arabic manuscript, Eves, p. 124. Page 131: Eves: custodians of much of, Eves, p. 178.

Chapter 6: Cardano and the Solution of the Cubic

Pages 136: excerpt from Gerolamo Cardano, Cardano, 1962, p. 2. Page 137: first excerpt from Cardano, Cardano, 1962, p. 25; second excerpt from Cardano, Cardano, 1962, p. 90; third excerpt from Cardano, Cardano, 1962, p. 240. Page 138: Oystein Ore: In the face of such tales, Ore, 1965, p. 33. Page 138: excerpt from Cardano, Cardano, 1962, p. 73. Page 138: Cardano: . . . when the desire, Cardano, 1962, p. 190. Page 140: excerpt from Cardano, Ore, 1965, p. 77. Page 141: excerpt from Cardano, Cardano, 1968, p. 96. Page 141: excerpt from Ludovico Ferrari, Fauvel and Gray, p. 259. Page 142: Tartaglia may have been fortunate, Eves, p. 200. Page 142: excerpt from Cardano, Cardano, 1968, p. 98. Pages 142–145: Cardano's solution of the cubic, Cardano, 1968, Chapter XI. Page 151: excerpt from Cardano, Cardano, 1968, p. 237. Page 153: Abel's proof, Smith, pp. 261–266.

Chapter 7: A Gem from Isaac Newton

Page 159: excerpt from Pierre de Fermat, Fermat, p. 291. Page 162: Isaac Newton: having uncleane thoughts, Westfall, p. 77. Page 162: excerpt from Newton, Westfall, p. 95. Page 164: excerpt from John Maynard Keynes, Fauvel and Gray, p. 421. Page 164 and 165: excerpts from Newton, Westfall, p. 143. Page 165: excerpt from R. S. Westfall, Westfall, pp. 137–138. Page 166: Viète likewise ran through, Viète, pp. 39–42. Page 167: Newton's version of his binomial, Fauvel and Gray, p. 403. Page 171: Isaac Barrow: a friend of mine, Whiteside, 1964, p. xii. Pages 174–176: Newton's approximation of π, Whiteside, pp. 100–101. Page 177: Newton: I am ashamed to tell you, Beckmann, p. 142. Page 178: Barrow: . . . a fellow of our College, Westfall, p. 202. Page 178: . . . so few went to hear Him, Westfall, p. 209. Page 178: excerpt from Humphrey Newton, Westfall, p. 192. Page 180: excerpt from Pierre-Simon Laplace, Burton, p. 383. Page 182: poems from Pope and Wordsworth, Fauvel and Gray, pp. 415–416.

Chapter 8: The Bernoullis and the Harmonic Series

Page 187: The paper carried the lengthy title, Fauvel and Gray, pp. 428–434. Page 190: only a trusted servant attended the funeral, Eves, p. 309. Page 192: excerpt from Johann Bernoulli, Kline, p. 473. Pages 196–198: Bernoulli's divergence proof, Dunham, p. 21. Page 198: Jakob Bernoulli: The sum of an infinite, Dunham, p. 22. Page 198–199: Bernoulli's poem, Smith, p. 271. Page 199–201: excerpts regarding Johann Bernoulli's challenge: Smith, pp. 645–648. Page 201: excerpt from Catherine Conduitt, Westfall, p. 582. Page 201: Newton: I do not love to be teezed, Westfall, p. 582. Page 201: Johann Bernoulli: . . . you will be petrified, Smith, p. 649. Page 202: excerpt from Nicole Oresme, Oresme, p. 76. Page 206: Jakob Bernoulli: If anyone finds, Dunham, p. 23.

Chapter 9: The Extraordinary Sums of Leonhard Euler

Page 211: Condorcet: He preferred instructing, Alexanderson, p. 276. Page 211: a publication backlog is reported, Eves, p. 328. Page 211: . . . there is ample precedent, Alexanderson, p. 316. Page 212: Euler: . . . quite unexpectedly I have found, Weil, p. 261. Pages 215–217: Euler's summation formula, Euler, Vol. 14, pp. 83–85.

Chapter 10: A Sampler of Euler's Number Theory

Page 224: André Weil: a substantial part of Euler's work, Weil, p. 170. Page 225: Pierre de Fermat: I would send you the demonstration, Ore, 1988, p. 272. Page 229: Christian Goldbach: Is Fermat's observation known to you, Weil, p. 172. Page 230–234: Euler's refutation of Fermat's conjecture, Euler, Vol. 2, pp. 68–74. Page 235: As of 1988, mathematicians know, Young and Bell, 1988. Page 236: excerpt from Carl Friedrich Gauss, Fauvel and Gray, p. 492. Page 239: Euler's "proof" exhibited, Struik, p. 99. Page 239: Leonhard Euler: The equation which determines, Euler, Vol. 6, p. 103. Page 239: Gauss: the clarity which is required, Fauvel and Gray, p. 491. Page 242: excerpt from Sophie Germain, Fauvel and Gray, p. 497. Page 242: excerpt from Gauss, Fauvel and Gray, p. 497.

Chapter 11: The Non-Denumerability of the Continuum

Page 247: excerpt from Morris Kline, Kline, p. 1157. Page 248: excerpt from Newton, Struik, p. 299. Page 249: excerpt from Leibniz, Edwards, p. 264. Page 249: excerpt from Berkeley, Smith, p. 630. Page 250: excerpt from Bishop Berkeley, Smith, p. 633. Page 250: excerpt from Augustin-Louis Cauchy, Grattan-Guinness, p. 109. Page 253: excerpt from Georg Cantor, Cantor, p. 86. Page 254: excerpt from Gauss, Dauben, p. 120. Pages 259–261: Cantor's non-denumerability proof, Fauvel and Gray, pp. 579–580. Page 261: Cantor: . . . remarkable . . . because of its, Dauben, p. 166. Page 265: He first proved, Fauvel and Gray, p. 579. Page 266: excerpt from Eric Temple Bell, Bell, p. 569.

Chapter 12: Cantor and the Transfinite Realm

Page 273: Cantor: I see it but, Grattan-Guinness, p. 187. Pages 274–276: proof of Cantor's theorem, Dauben, pp. 165–167. Page 277: excerpt from Cantor, Dauben, p. 166. Page 278: Joseph Dauben: not only as God's messenger, Dauben, p. 147. Page 278: excerpt from Cantor, Dauben, p. 147. Page 280: excerpt from Cantor, Gillispie, Vol. 3, p. 57. Page 281: David Hilbert: No one will expel, Kline, p. 1003. Page 282: Hilbert: . . . very plausible theorem, Calinger, p. 661. Page 283: excerpt from Cantor, Dauben, p. 298.

References

Alexanderson, G. L. "Ars Expositionis: Euler as Writer and Teacher." *Mathematics Magazine* 56 (November, 1983) 274–278.

Beckmann, Petr. *A History of π*. New York: St. Martin's Press, 1971.

Bell, Eric Temple. *Men of Mathematics*. New York: Simon & Schuster, 1937.

Bernoulli, Jakob. *Ars Conjectandi*. Basel, 1713.

Borwein, Jonathan M. & Borwein, Peter B. "Ramanujan and Pi." *Scientific American* 258 (February, 1988) 112–117.

Burton, David M. *A History of Mathematics: An Introduction*. Boston: Allyn & Bacon, 1985.

Calinger, Ronald. *Classics of Mathematics*. Oak Park, Ill.: Moore Publishing Co., 1982.

Cantor, Georg. *Contributions to the Founding of the Theory of Transfinite Numbers* (Trans. by Philip E. B. Jourdain). New York: Dover (Reprint), 1955.

Cardano, Girolamo (a.k.a. Jerome Cardan). *De Vita Propria Liber* (Trans. by Jean Stoner). New York: Dover (Reprint), 1962.

———. *The Great Art, or The Rules of Algebra* (Trans. by T. Richard Witmer). Cambridge, Mass.: M.I.T. Press, 1968.

———. *Opera Omnia* (1662 edition), Vol. IV. New York: Johnson Reprint Corporation, 1967.

Cicero. *Tusculanes*. Paris: Societe d'Edition "Les Belles Lettres," 1968.

Dauben, Joseph. *Georg Cantor: His Mathematics and Philosophy of the Infinite*. Cambridge, Mass.: Harvard University Press, 1979.

Dunham, William. "The Bernoullis and the Harmonic Series." *The College Mathematics Journal* 18 (January, 1987) 18–23.

Edwards, C.H. Jr. *The Historical Development of the Calculus*. New York: Springer-Verlag, 1979.

Euler, Leonhard. *Opera Omnia* (1), Vols. 2, 6, and 14. Leipzig, 1915–1925.

Eves, Howard. *An Introduction to the History of Mathematics*, 5th Ed. Philadelphia: Saunders, 1983.

Fauvel, John & Gray, Jeremy. *The History of Mathematics: A Reader*. London: The Open University, 1987.

Fermat, Pierre de. *Oeuvres*, Vol. 1. Paris: Gauthier-Villars et Fils, 1891.

Gillispie, Charles C. (Ed.-in-chief). *Dictionary of Scientific Biography* (in 16 vols.). New York: Scribners, 1970.

Grattan-Guinness, Ivor (Ed.). *From the Calculus to Set Theory, 1630–1910*. London: Duckworth, 1980.

Hall, Rupert. *Philosophers at War*. New York: Cambridge University Press, 1980.

Hardy, G.H. *A Mathematician's Apology*. Cambridge: Cambridge University Press, 1967.

———. *Ramanujan*. New York: Chelsea, 1959.

Heath, Sir Thomas L. *A History of Greek Mathematics* (2 Vols.). New York: Dover (Reprint), 1981.

———. *The Thirteen Books of Euclid's Elements* (3 Vols.) New York: Dover (Reprint), 1956.

——— *The Works of Archimedes*. New York: Dover (Reprint), 1953.

Hilbert, David. *Foundations of Geometry* (Trans. by E. J. Townsend). Chicago: Open Court, 1902.

Kline, Morris. *Mathematical Thought from Ancient to Modern Times*. New York: Oxford University Press, 1972.

Loomis, Elisha Scott. *The Pythagorean Proposition*. Washington: National Council of Teachers of Mathematics, 1968.

Ore, Oystein. *Cardano: The Gambling Scholar*. New York: Dover (Reprint), 1965.

———. *Number Theory and its History*. New York: Dover (Reprint), 1988.

Oresme, Nicole. *Quaestiones super Geometriam Euclidis* (Commentary by H.L.L. Busard). Leiden: E. J. Brill, 1961.

Plutarch. *The Lives of the Noble Grecians and Romans* (trans. by John Dryden). New York: Modern Library, no date.

Proclus. *A Commentary on the first book of Euclid's Elements* (trans. by Glenn R. Morrow). Princeton: Princeton University Press, 1970.

Rickey, V. Frederick. "Isaac Newton: Man, Myth, and Mathematics." *The College Mathematics Journal* 18 (November, 1987) 362–389.

Russell, Bertrand. *The Autobiography of Bertrand Russell* (2 vols.). Boston: Little, Brown & Co., 1951.

Sandburg, Carl. *Abraham Lincoln: The Prairie Years*. New York, Harcourt, Brace & Co. 1926.

Smith, David E. *A Source Book in Mathematics*. New York: Dover (Reprint), 1959.

Struik, Dirk. *A Source Book in Mathematics: 1200–1800*. Princeton: Princeton University Press, 1986.

Trudeau, Richard. *The Non-Euclidean Revolution*. Boston: Birkhauser, 1987.

Viète, Francois. *The Analytic Art* (trans. by T. Richard Witmer). Kent, Ohio: Kent State University Press, 1983.

Weil, André. *Number Theory: An Approach through History*. Boston: Birkhauser, 1984.

Westfall, Richard S. *Never at Rest: A Biography of Isaac Newton*. Cambridge: Cambridge University Press, 1980.

Whiteside, D. T. *The Mathematical Works of Isaac Newton*, Vol. 1. New York: Johnson Reprint, 1964.

Wolfe, Harold. *Introduction to Non-Euclidean Geometry*. New York: Dryden Press, 1945.

Young, Jeff & Bell, Duncan A. "The Twentieth Fermat Number is Composite." *Mathematics of Computation* 50 (January, 1988) 261–263.

Index